GROWING UP AUTOMOTIVE

A Book for the Aspiring Young Technician and the Unaware Consumer

BY CHARLES ROSE

This book is the sole copyright of the author. It is an autobiography of the author's experiences. It cannot be reproduced in whole or part digitally or in print by any means without the express consent of the author. To obtain the author's consent write to the publisher:

editors@emerald-design.co

Summary

Having spent the majority of his life immersed in the automotive world, Charlie Rose shares his most memorable moments.

BISAC Categories

TEC009090 Technology & Engineering > Automotive

BIO026000 Biography & Autobiography > Personal Memoirs

ISBN: 9798568488668

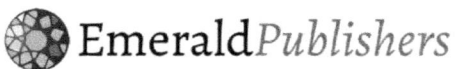

DEDICATION

This book is dedicated to all former and current automotive technicians around the world who work tirelessly to keep the population mobile.

I would also like to extend my appreciation to the hard-working men and women throughout the transportation industry. The truck, bus, train, marine, motorcycle, aviation, and industrial technicians are all dedicated to keeping the world supply chains moving.

Without the wrenches turning, the wheels certainly would not.

CONTENTS

Introduction	1
Chapter 1: The Early Years	5
Chapter 2: Changing Jobs	25
Chapter 3: "My Mechanic"	42
Chapter 4: Mechanic vs. Technician	46
Chapter 5: Technology	59
Chapter 6: Self-Employed	67
Chapter 7: Fire and Safety	72
Chapter 8: Demographics	82
Chapter 9: Tools	85
Chapter 10: Teaching Automotive	89
Chapter 11: Coworkers and the Shops	102
Conclusion:	115
About the Author	118

INTRODUCTION

When I began writing this book, I had been looking for a summer reading assignment for my automotive students that was not a how-to manual, but a comprehensive idea of what the automotive trade looks like from someone who's lived it. Searching bookstores and online yielded very little, so I began typing about my own experiences in the automotive industry. The purpose of this book is to provide insight into my life and career as an automotive technician in order to inspire others and educate consumers. I will take you from my early days of fixing automobiles in the '70s right up to today, including how I transitioned into teaching automotive to high school students bound for the trade. Most automotive technicians I have spoken to or worked with had stories and experiences similar to mine. Some chose to remain at one location throughout their career, while others (like myself) have changed jobs several times. I have worked in over 25 automotive repair establishments in my lengthy career, including being self-employed briefly. All of my job changes were of my own choosing, and I was never "let go" from a job. I spent the majority of my career on the cutting edge of technology where the factory help desk and manufacturing engineers served as a key resource for me. In this book I bring you into the service department of various dealerships, independent garages, gas stations, engine and transmission rebuilding shops, tire stores, chain stores and my time being self-employed. I will introduce you to the people that I have worked with and for, as well as the honesty and dishonesty that I have witnessed. There are humorous stories of my experiences as well as sad ones. Real names may or may not be used, but all situations and stories are fact. In reading this you will begin to understand why there is a lot of distrust within the automotive repair industry by consumers, and why searching for a repair facility that can be trusted is critical. I look to give those seeking a career in the automotive repair industry a little more insight into the realities of life in a shop environment. This book will also give consumers a "heads up" when seeking auto repair. I hope you enjoy reading this book as much as I enjoyed reliving my career and the career of others.

It is amazing to me that the most basic automobiles had come into being just one hundred years before I began my career. It was back in the year 1870 when German inventor Siegfried Marcus built the first gasoline-powered combustion engine, which he placed in a push cart. He then went on to build four progressively more sophisticated combustion engine cars over a ten-to-fifteen-year span. Nikolaus August Otto was patenting the world's first four-stroke engines in Germany in 1885. Back in the United States, with the Industrial Revolution well underway, demand for this technology was growing. In the early 1900s, such visionaries as Henry Ford and the Dodge brothers were working to bring this new product to life so every household could afford one. The original Model T, released in 1908, packed a 2.9-liter four-cylinder engine with just twenty-two horsepower. Roads and bridges were being built to accommodate the evolution of the "automobile." To promote the automobile, newspapers sponsored an 87-kilometer race from Chicago to Evanston on Thanksgiving Day in 1895. The average speed of this race was 24.15 kilometers per hour (15 mph). There was a large turnout; interest in the automobile was growing. As the demand for this new way of transportation soared, so did its technological advancements. Tremendous changes had been made since that time as well as many various body designs and engine configurations. It seems that each decade brought a new version of the automobile. However, the decades since the 1970s—when I began my career—the greatest advancements have unfolded. Front wheel drive, hybrids, standardized fuel injection, antilock brakes, and four-speed electronic transmissions are some of the systems I have witnessed from their first prototypes . Those early systems progressed to eight- and nine-speed transmissions, electric vehicles, electronic stability control, and lane departure systems, all the way to self-driving vehicles. I bring up the history of automobiles because mechanics of the 1950s had no idea that they would be working on automatic transmissions, disk brakes, and power steering systems, just as I had no idea I'd be working on the systems I mentioned. I always tell my students that they will become experts on systems that have not even been invented yet. I let them know they may play a part in bringing some new technology to market.

Here are my stories from this era of change.

CHAPTER 1:

THE EARLY YEARS

I remember sitting in my eighth grade classroom being asked which path I would like to take in high school and beyond. My options were regular high school, which I believe was explained to me as a business program, or vocational school. I had no idea what either meant. And I had no idea that I actually had to choose what I wanted to be for the rest of my life at such a young age, or so I felt. After being sheltered in a private "saint" school for the previous eight years, reality hit me hard. Coming from a family of nine, which was the norm in those days, I felt I had resources at home that would help me.

"Dad, what should I be when I grow up?" That was my approach at the time. Tinkering with whatever I could find around the house was a passion of mine as far back as I can remember. I grew up with Lincoln Logs and red-brick building sets, which were the precursor to Lego. I also had an Erector Set, a kit with nuts and bolts, wheels, gears, and metal brackets. I had electronics building kits, where I could build radios and the like and I also had multiple electric train and car sets. Back in eighth grade, I asked Dad what he thought of my path. Knowing my tinkering abilities, Dad felt I would find my niche in one of the trades. "What's a trade, Dad?" was my response. Dad, a blue collar worker himself, explained plumbing, electrical, and automotive—I don't remember hearing about any others. I heard "fixing cars" and a light went on. *That's it! That's what I want to do.*

The following week, I met with the high school guidance counselor, who tried to fill my head with doubt. He was talking about being dirty all the time and having to tow vehicles in foul weather. I was undeterred. "Sign me up!" I remember saying. The following year, I began my pre-automotive course. The first half of my high school freshman year included small engine repair mixed with electronics. I was disassembling small engines and learning from my teachers what the various parts are and what they do. In electrical, I was soldering and learning about resistors and how to read them. I was a natural and easily transitioned into the auto shop for the second half of my first year. The instructors at the time had newcomers like me file a block of steel until it had perfect corners and flat sides. This was all done by hand. I remember thinking *How is this going to help me fix cars?* I now understand that it was to get us used to hand tools. The rest of the year went well, and I split my time between the shop and the classroom. In the shop, we quickly moved on to performing all kinds of repairs with instructor supervision. We were doing brake jobs, tune-ups, oil changes, and alignments, and in some cases we went as far as rebuilding engines. Sometime during my sophomore year, the Environmental Protection Agency, the EPA, had realized that brake and clutch dust can be detrimental to mechanics' health. The brake industry came to our school to educate us on the dangers of asbestos. Prior to this we would all just take compressed air and blow the dangerous brake dust into the air around us. We worked and breathed in this environment. This was the norm for cleaning brakes back then, and nobody thought anything of it. Luckily, I'm not one of the thousands from around the world who has suffered the effects of asbestos exposure.

In the next two years of my high school career, I had a great experience and I learned quite a bit. I was not a great student and barely maintained a C average (it actually may have been a bit lower) due to the classroom work, but I excelled with the hands-on portion. Many of us in the trades didn't consider ourselves to be cut out for schoolwork but loved using our hands to make a living. We were considered "vokies" by the regular high school students, which implied that we were not that smart. I would like to compare how the take-home pay differed between the "vokies" and the regular high school students over the years. My guess would be that the trades paid more, as many of us went on to start our own successful businesses and bypassed college debt.

I made great friends in high school and retain many of them to this day. Even while I was still in high school, I began to repair my own car, as well as those of my family and friends. I was not afraid to tackle any job and would always have some project going on in my backyard. I remember a cold winter day when I removed a transmission from my car lying on my back and I could barely feel my fingers. When I went inside, I ran my hands under cold water to try to thaw them out. Even though the water was cold it still felt like they were burning. I was only focused on one thing—getting the job done whatever the cost. *Ah, to be young again!*

Growing Up Automotive

In high school years and for many years after, I was into modifying cars to make them faster. I have been through many fast cars in my time, and I've always had one of the fastest in the area. The one car that stands out was a 1968 Pontiac Firebird that I purchased without an engine. I still owned a 1969 Grand Prix with a 455-cubic-inch engine with three two-barrel carburetors, also known as a six pack or tri-power. This Grand Prix had been hit while parked in front of my house. The insurance company declared the vehicle a total loss. I bought the vehicle back because of its personal and sentimental value. I had recently rebuilt the Grand Prix engine with some high-performance modifications. I placed that engine into my recently purchased Firebird and had myself one powerful automobile. I ran just under twelve seconds in the quarter mile at a local drag strip and, to me, that was plenty powerful. Cars had become part of my identity, and motor oil was running through my veins.

After graduation, I was up and running in my newfound career. I took a job at a local gas station pumping gas and fixing automobiles with a big burly biker named Jay who took me under his wing. It didn't take long for me to have the confidence to service brakes, timing chains, tune-ups, exhaust systems, front-end work, batteries, and just about any other job that was thrown my way. Jay, my mentor at the gas station, really helped me advance my career quickly. He also had a big old Harley Davidson that was loud and ratty-looking. I was a scrawny kid at the time, but when he told me to take the Harley for a ride, I was all in. When I returned about an hour later, I was all smiles. That was a loud, fun bike, and he knew I would enjoy it.

One story that stands out (and still kind of makes me cringe) is the night that a "friend" stole a car and drove it into the woods. Friends had taken the battery out by cutting the cables, stole the tires, and took the radio and anything else that was easy to access. This was at a time when stealing cars was fairly easy and was being done what seems like everywhere, so we really didn't think too much of it. Well, the next day I went to work and, like most days, I grabbed a cup of tea and waited for my first job. It was around mid-morning, and what I saw pulling up on a tow truck had my heart pounding. It was the car that had been stripped of all its parts the night before. With my heart still pounding and my hands shaking, I asked my partner what that car was doing here. He went on to tell me the story of how someone had stolen it that prior evening and that the owner had come to install new tires, battery, battery cables, and a radio. The ignition was also broken and needed several parts. I was relieved that no one knew I had something to do with it, but I could not believe the chances that it came to my particular gas station. I knew where the original parts were—not far from the station—and I was plagued with guilt. There I was selling this customer brand new parts at full price and convincing myself that insurance would somehow make it okay. Well, maybe it did (a little), but I could not get rid of those parts fast enough. After completing the repairs, I made sure I was nowhere to be found when the customer came to pick up his vehicle. I learned my lesson that day and stepped away from that unethical practice.

I moved on from the gas station, which was a real confidence builder for my abilities as a young mechanic and a true learning experience. My next stop was changing oil at a Buick dealership. I got fast at that task and really began to take pride in everything I did, but I wanted more. Just changing oil and taking care of small items was not allowing me to grow as a technician as I wanted to. I had been doing much more at the gas station before I came to the dealer and wanted to get back to larger repairs. Dealers at the time had specialists in every area throughout the shop so moving up meant specializing in a certain area, which wasn't an option at that stage of my young career.

I made some good friends at the dealership, and almost every Friday we shared what became affectionately known as a "liquid lunch"—when the bar tab was larger than the food tab. I'm not proud of that, but it was a different time. I'll never forget a little old lady who would come into the dealership religiously for her oil changes and maintenance and had no idea what she was driving. This woman owned an immaculate, at the time twenty-year-old, low mileage Buick Riviera that had been her late husband's. The Riviera had the big 455ci engine with two four-barrel carburetors and all the available options. Everyone in the shop wanted to buy the car from her, but she was happy just putting along in her collector's dream car. I always had a smile on my face when I watched her drive away feathering the throttle and obeying the speed limits. I remember another time when I had one oil change going on the lift and another vehicle parked behind by the door waiting to have the oil changed as well. When I went out to bring in the second vehicle, it was gone. After asking around to see if someone had moved the car, I realized that it was stolen right out from under my nose. It was a new, fully loaded Buick LeSabre that I now had to tell my manager was missing. Shortly after I told him, I saw the owner of the dealership, my manager, another person I assumed to be the owner of the vehicle, and three other gentlemen dressed as nicely as the owner of the vehicle approaching me. They came over to ask a question about the missing vehicle with a grin on their faces and walked away. *Very peculiar*, I thought, *for someone who just had a car stolen.* Well, rumor had it that the owner was well connected and had the vehicle "taken care of." Was there a story behind the vehicle—like a body in the trunk or something? I find it hard to believe that he just wanted to replace it by having it stolen and not trade it in, especially when he seemed to have a great relationship with the owner. I guess I'll never know.

At that time, my dad knew a lot of people. For my next job, my dad had me interview with an engine rebuilding company. The owners were Portuguese like us, and he knew them through mutual friends. When I first walked in, I was overwhelmed and excited about all that was happening inside that nondescript building. Down at one end of the building mechanics were taking out and installing engines in a wide variety of vehicles. There were trucks, cars, front end loaders, backhoes, sports cars, luxury cars, and vans. You name it, it was there. They turned nothing away. The center of the building was where all the repairs were performed, and that's where I spent a majority of my time. The rest of the building was dedicated to rebuilding engines. Those guys were my heroes. There were giant tanks that used steam for cleaning, machines for cutting and grinding, lathes for precision work, and a wide variety of measuring equipment. They had benches laid out that made the area look more like a science lab than an automotive engine rebuilding company. It was primarily my responsibility to fine tune the vehicles after an engine was installed, or diagnose them to determine if a vehicle actually needed an engine or not. I remember one vehicle that was towed in because someone told the owner he had thrown a rod. In case you don't know, throwing a rod is breaking a connecting rod that is attached to a piston deep inside the engine. This makes a tremendous noise and in most cases, the engine won't run and it's in need of a full replacement. This vehicle was running, and not too badly. Upon further investigation, I noticed that a push rod, not a connecting rod, had come up through the valve cover. This was a much cheaper repair than the customer expected, and we looked like heroes. Another vehicle was blowing white smoke out of the tailpipe after an engine rebuild. When we see this in the field, it is often due to burning coolant in the engine because of a defective head gasket or something similar. They called me over to inspect the vehicle and upon mapping out all the vacuum lines, I found the windshield washer tank hose connected directly to the intake manifold. This, as you probably guessed, was allowing windshield washer fluid to be burned directly in the engine. Problem solved! I also remember a van towed in that had oil coating inside the passenger's window and seating area. It was a mess! The engine oil had filled with gasoline from a flooding condition due to carburation. When the owner of the vehicle went to start the van, the fumes in the oil exploded and blew the valve cover completely off the engine. I didn't notice, but the driver's seat probably needed to be cleaned as well.

Chad, a co-worker who was about my age, became a good friend, and we partnered up on many projects. One day, we were asked to bring back a large front end loader that had broken down a couple of blocks away. When I say large, I mean the one that swivels in the middle and you can almost walk under—that big! We really had only one option when it came to hauling heavy equipment: the converted white cab-over. This was a truck that was designed to be the front half of an eighteen wheeler. This truck was not very successfully converted into a tow truck. The weight distribution was all wrong, something I will talk about later in the book. Chad and I took this large wrecker and went to see how we could get this monstrous front end loader back to the shop. Being young and fearless, the only option we saw was to strap a chain from the wrecker to the loader and pull it back to the shop. This seemed like our best chance. I guess we could have called an outside company to tow it properly, but we decided to make it happen somehow. I climbed up into this mountain of a machine and figured out how to release the parking brake. I yelled ahead to Chad that I was ready for our little adventure. We were laughing when we realized that it might not go too smoothly. We started moving, and I was steering—or trying to steer—the loader. With the engine off and no hydraulic pressure to run the steering system, it becomes almost impossible to head in the direction you intend. But nothing can stop two young men that set a goal for themselves. "It's only about a quarter mile. What could go wrong?" So here we were, this big semi-truck converted to a tow truck with a chain latched onto this even bigger front-end loader as we came into our first corner.

"I don't think they liked those bushes anyway, keep going!" I yelled as we tried to round our first corner. *Okay, this ought to be good*, I thought to myself as the next corner with a fence was approaching. *I'm not gonna make it! I'm not gonna make it!* Well, I didn't make it and kind of chuckled as I ran over the four-foot stockade fence that perhaps someone was thinking of replacing anyway. That didn't work out so well. "Maybe we should slow down!" I yelled to Chad as we tried to round yet another corner. I doubt he'd heard me because I know there were some bicycles and assorted toys in that last yard that may need to be dug out of the earth to be used again. When we looked back at our route, it was more of a straight line to the pickup site than the zigzag a map would suggest. That was sure an interesting adventure that I would probably not repeat should the occasion arise.

Every one of us that towed for that company had a funny story about a time we needed to tow a heavy truck. Our "tow truck," a converted white cab-over, had a flat nose from the windshield to the bumper that forced the driver to look straight down at the front of the truck. It didn't have the proper weight distribution to be towing anything heavy. One day, I found this out the hard way when I was towing a long rack truck full of tires from a location just south of Boston. After I hooked up to the truck, I tried to pull out of the parking lot. I said *tried* because when I tried to turn the steering wheel, nothing happened. The weight in the back of the vehicle in tow would raise the front end of the tow truck so much that the front tires barely touched the ground. The tow truck actually pivoted on the rear wheels like a seesaw. Pressing on the accelerator only made things worse, and the front end would rise until the towed vehicle touched the ground, basically popping a wheelie. At that point, I had to have the company remove at least half the tires from the back of the truck I was towing to lighten the load, so I could drive somewhat safely. I was finally able to leave for our engine rebuilding shop on the other side of Boston about twenty miles away. All was going well until I was in traffic coming up a hill on the three-lane highway leading out of Boston. Every time I stepped on the gas, the whole nose of the truck, with me in it, would ascend skyward until the front wheels of the truck I was towing contacted the ground. I could look down at the tops of the cars around me. The steering was inoperable when the front wheels were in the air. That was a nerve-wracking ride that may have taken a couple of years from my life. It's interesting to note that I began to have much more room around me after the first loss of gravity.

Another tow with that same white cab-over left me shaking and thanking the good Lord that I'm still alive. I was instructed to bring a broken down vehicle from Route 2, a busy, narrow two-lane highway back to our shop. I had just finished hooking the car to the tow truck in the narrow breakdown lane of this two-lane highway. After completing the attachment to the car, I pulled myself up into the truck using the inside door handle. I was moving as fast as I could because the traffic was real close. I was sitting in the driver's seat after jumping in, and my left hand was on the outside edge of the door pulling it in to close. The instant I transferred my hand from the edge of the door to the inside door handle to pull the door closed, a big U-Haul box truck swerved too close to me. The mirror from that truck came in contact with my door exactly where my hand, and my body, had just been a fraction of a second prior. The noise was incredible! The truck clipped me with such force it bent my door and ripped the mirror from the U-Haul and the mirror went flipping down the highway after impact. The contact made such a loud bang that my left ear was ringing after the incident. The driver didn't even slow down, let alone stop. It took a minute for me to calm down, but I knew I needed to get moving to prevent any other problems. After that incident, I went to all the U-Haul dealers in the area searching for that truck with no luck. I assumed that they were just passing through on their way somewhere else. The driver had no idea if they had hurt or even killed me. How could they have not understood? Perhaps drugs or alcohol were involved—I will never know. I was certainly more aware of the danger of towing a vehicle from the side of the road after that incident. The current law of "move over and slow down for emergency vehicles" that is posted on the side of the roadways needs to be taken seriously. This is a problem that occurs more often than most people know. Police that stop vehicles on the highway fall victim to this quite a bit. I have a family member who was a state trooper in the 70s and was struck by a vehicle while on a routine stop on the highway. After many operations, he walks with a cane and has never been able to return to his job. That is just one of many stories out there.

I worked with some wonderful people at that engine rebuilding shop. Delphine, my immediate supervisor and mentor, was a soft-spoken Portuguese gentleman. I learned quite a bit from him in my three years there. There was a time when the oxygen hose exploded off of the oxygen/acetylene torch set that produced a one-foot flame backed by two thousand pounds of pressure. Several of us began to run toward the exit, unsure of what danger lay ahead after hearing that explosion. Delphine saw what had happened, walked over to the source of the problem and simply shut off the valve without any emotion. I looked at him in disbelief. In my head I was thinking *Rock star!*

My abilities as a technician really blossomed there because of my experiences and meeting many good technicians. There was one gentleman there who always worked alone. One day early on, I walked over and began a conversation with him. I noticed he was very neat and organized. His hair was perfectly combed, the floor around him was clean, as was his toolbox and bench area. We called him Sam, which I'm sure was short for a much longer Portuguese name. Sam did lots of heavy work like removing and installing engines and transmissions in just about anything. I thought to myself, How is it possible to do this kind of work and remain so clean and organized? He was just so professional in every way, and he knew how to capture or drain any fluid from the vehicle before it ended up on the floor. Sam is still in my head when I do work today. I have been inspired by his work ethic throughout my career. I make sure the fluids are drained properly when disassembling fluid-filled components and try to only use force when it's necessary. I've noticed that automotive technicians model themselves on their experiences working with other technicians or instructors. I try to be that role model for the students that come through my automotive program. Instead of forcing parts together or apart, I would rather see them be patient and finesse the parts that may be a bit stubborn. This always is the best route to take.

I began to notice all the health hazards I was exposed to in that engine rebuilding company. Every bay had a half-full pail of gasoline for washing parts in. That's correct—open containers of gasoline dispersed throughout the shop. One bay in particular had live exposed electrical wires hanging from the ceiling. When a car was raised, sparks would come off the body or mirror of the vehicle. Many cars left that shop with burn marks. Another major health concern was the lack of vehicle exhaust ventilation for the building. Whenever a vehicle starts up after receiving a new engine, it smokes pretty badly at first. You can imagine what the air was like in there most days. Clouds of smoke would form and hang around for a while. I knew the hazards of carbon monoxide and was not happy with the situation. I had been taught that a running vehicle in a closed one-car garage could build up enough poisonous carbon monoxide to kill a human within three minutes. While this building was much larger than a closed one-car garage, the amount of exhaust was exponentially higher than that of one vehicle. In the summer months, the doors would be open, so I justified that it was not so bad. During the winter months, we always toed the line between breathing and freezing by opening the overhead doors when we felt we were reaching a dangerous level of poisonous gasses. I would talk about these things with my family daily when I got home from work. My sister, who happened to work for the Occupational Safety and Health Administration (OSHA) at the time, overheard me talking about this one day. In case you don't know, OSHA is the government watchdog agency for employee safety. Well, all my sister needed to hear was that her baby brother was exposed to these kinds of workplace safety hazards. The next day, I was contacted by an agent from her office and told him the things that I knew about. Within a week, the agent and his team came in for a look around the shop. The inspector informed me of the many violations that they found. I'm not sure how my workplace found out I was the one who informed OSHA of their violations, but I was let go of my job the following day. After that happened, I was informed by the OSHA field agent that by law I could not be fired for reporting safety hazards. The employer would have to reinstate my job and make up for any missed pay, or pay me for as long as I was out of work. It was killing me to go find another job, but I certainly was not going to be able to work for them again. I stayed out of work as long as I could. Two weeks was all I could tolerate because it just wasn't in my nature to not work. In the end, I was paid for all the lost time and was offered my job back. The shop owners were fined and were required to eliminate the hazards. I felt good knowing that maybe I saved some injuries. Looking back at that experience with OSHA and workplace safety, I would not hesitate to do it again. This time I would not wait for an injury to occur.

These early years really defined me as an aspiring young technician. I gained a lot of knowledge, experience (some good, some not so good), and insight into the world of an automotive technician. I grew up automotive, and I was well on my way.

CHAPTER 2:

CHANGING JOBS

I landed my next job at a new car dealership, which turned out to be the start of a thirty-plus-year career with Chrysler Corporation. It was a small mom-and-pop dealership owned by three brothers who each ran different aspects of the operation. One was the main financial operations director who was also on the board of a local bank. Another ran the showroom sales force, while the third ran the service department. They were all very nice and treated me well. Lee Iacocca ran Chrysler in those days and was doing great things. I learned a lot while working there, and I was constantly being sent to the factory for training. I performed lots of recalls and warranty work. I also worked with factory representatives who would be sent into the field if the shop needed extra help on a repair. I learned about the pay system used at many dealers at that time.

This particular dealer had a unique pay system allowing for an increase in hourly wage based on the dollar amount you made for the company each week. The more you made for the company, the more your take-home pay. This is the norm when working in the auto repair industry, the faster and more productive you are, the more you make. This comes at a cost though. When working fast and keeping your eye on the prize, it's possible to make more mistakes, and too many mistakes are costly to customer satisfaction and job security.

The three brothers who owned the dealership would never allow a *red* car to remain on the lot due to superstitious reasons. If somebody wanted to purchase a red car, the owners would order it but the customer needed to pick it up the day it was delivered—it could not stay on the lot. I thought that red car superstition was quite interesting. Some years later, I was watching a movie called *Used Cars*. A used car dealer advertised that he had a mile of used cars. The local authorities questioned his advertising slogan and measured all of his inventory end to end. He was short of a mile and had to produce the full mile of used cars within a week. He began buying cars around the country and hired drivers to bring them to his lot. One driver had the same superstition about red vehicles and was given a blue car to drive. The paint began to peel on the blue car, exposing the red paint beneath. When the driver realized he was driving a red car, he immediately hit the brakes and climbed out and refused to continue. An online search also shows that red cars have a seven percent greater chance of being involved in an accident as compared to other colors. Maybe the dealer superstition isn't as far-fetched as I had thought.

The dealership was full of unique characters. One such character, Mark, was always talking to himself. You could walk by his bay and hear him having a full-blown conversation. I would look around and there was no one in sight. Well, isn't this interesting, I would think to myself. This same gentleman was a wild story teller, and I found out he was drunk most days. At break, Mark would have story after story for us. One of his most popular ones was about an eighteen-foot python that he had as a pet. He would tell us how he would feed it chickens and rodents. He would also tell us how it would get out of its enclosure at times. Well, one day, he was telling us about how it escaped from its holding tank and he had to wrangle it back in. An older man, Joe, who was very quiet and kept to himself, snuck up behind Mark and started wrestling with an air hose as if it were a snake. He had the hose wrapped around himself as he went down to the floor like it was strangling him. Mark could not see Joe as he talked, but we all could. We were laughing so hard that Mark thought his story was a big hit. That was certainly unforgettable. Some of the employees from there who I see from time to time still recall this story some forty years later. Although we had fun at Mark's expense, it made for a pleasant working environment.

Another story I recall from that dealer was when we were looking for a water leak in a trunk. Most of the time, we would put someone in the trunk with a flashlight, shut the opening and spray it with a hose so they could try to locate the source of the leak. This method went well until it was time to get a new guy out of the trunk. "Where are the keys?" we asked Frank who was in the trunk. "I have them right here" was the response I received. After a pause and a moment of laughter, he said "Oh, I guess that wasn't a good idea, now was it?" I proceeded to have him remove a rubber drain plug from the trunk floor and drop the keys out to me. Frank still thinks about this whenever he's on the lookout for a leak.

That Chrysler dealership also seemed to be located in some kind of lightning zone. My coworkers told me about a time when lightning had hit the ground behind the shop and then traveled across the shop floor. The lightning bolt dissipated into the water pipes tucked in the corner, creating a loud rattling noise. There was a burn mark on the floor that traced its path. I'd witnessed a couple of strikes firsthand. One happened while I was in the shop walking behind a car with its hood open. When I was directly behind the car I saw a bright flash and heard a loud crack on the bench directly in front of the car. When I went to investigate, I realized that a lightning bolt had come through the open window and seared the bench. I was glad no one was standing there at the time. Another strike I witnessed happened while I was a block away from the dealership while road testing a customer's car. I was stopped at a red light when a massive bolt of lightning struck the transformer about ten feet from me and just within view. I saw smoke and what was probably the beginning of a fire. The stoplight had gone out and when it was safe I bolted back to the shop. When I got there, we had no electricity and my coworkers told me that a bolt of lightning had hit the roof of our building. I was sure glad that it wasn't me on the receiving end of the 100 million-plus volts from the heavens above.

While I was there, I'd become a problem solver. I loved figuring things out, and that really hasn't changed over the years. There is personal satisfaction in repairing a challenging problem. Working with problem automobiles allows for the flow of your creative juices, and it keeps you interested. I spent roughly three years working there and I felt it was time to move on.

Throughout my career, I have been a nomad in that I never stayed at one place for too long. I chased better pay and working conditions, or at times I needed to leave due to the way I saw customers and employees treated. I did learn that I'd never be without a job in the automotive industry as long as I had my health. There are plenty of "parts replacers" out there but few people that take the time to understand how things operate. My ability and deep understanding of vehicles has made me indispensable and allowed me to demand top pay. When I work on a vehicle, I don't focus on a dollar sign like so many in my field, but as the second largest investment that most people will make. I view customers' vehicles as what they are, providing valuable transportation and an important part of people's lives. Their owners are mothers, fathers, sisters, brothers, cousins, aunts, uncles, and friends. One of my highest priorities is keeping them safe on the road and keeping their businesses running.

For the next few years, I hopped around to different jobs. One stop was a transmission rebuilding shop where I worked removing and replacing transmissions. That was a tough job that I never really enjoyed. Pulling transmissions out of vehicles all day every day is not a job that can be done for a long duration. It will take a toll on you, both physically and mentally. The owners of the shop were good people, though. The father ran the business and the son worked the day-to-day operations. They appreciated their workers and compensated us fairly especially around Christmastime with dinners and bonuses. I kept in touch with them for many years afterwards. They treated me like family, giving me the best deals they could when a friend or relative needed transmission work. When their shop closed, I was sad.

My next stop was an independent brake and tire shop. This shop was interesting... to say the least. The father and son who owned it would fight all the time. One day they argued so much that we were all sent home and they closed the doors. The next day they opened as usual with no explanation. *Interesting,* I thought but went about my business. One particular story from that tire store sticks with me, and I still chuckle when I think about it. There was an "alignment pit" near the entrance to our shop. This was a hole in the ground that the vehicle was driven onto the ramps which extended into the center of the "pit". The technician walked down into this pit to perform the alignment. All of our front end alignments were performed here. Well, on this particular day, a customer was waiting for an alignment and we were quite busy. She was pacing by the railing near her car, which was in the alignment area. It was getting close to lunchtime, so my coworker mounted the alignment heads onto her wheels. He turned some dials and pressed some buttons as she watched. Well, my coworker goes to lunch and comes back some 45 minutes later, looks at the alignment screen and says to the customer who was still waiting, "It looks like it's all done, I had it on automatic." Just so you know, there's nothing automatic about alignments as all steering and suspension components must be adjusted manually if the specifications are off. We all laughed after that but those are things that an uneducated consumer faces. I don't consider the facility to be totally unethical, or I wouldn't have stayed, but they did have their moments and that was certainly one of them.

Most mechanics/technicians (and I will explain the difference in a later chapter), do not like interacting with customers. They would rather be given a vehicle and be left alone to fix it. In fact, service writers and managers think and treat customers like a burden. I've always had trouble working with people that have developed this mentality.

Do they not understand that the customer should be their best friend, as their pay depends on it? They only see the vehicle as a dollar sign and not someone's transportation. Most mechanics have grown up fixing vehicles and have never had to experience the other end of the repair process--the customer's perspective. If they had, things would be different. One Dodge dealership I worked at bothered me. The service manager, who was very grumpy, would greet and write up customers with his head down, grumbling as he spoke into the paper. As a customer faced with that situation, I would already be thinking negatively about the establishment. He just didn't understand that customer interaction is the most important part of any business. Shame on the owner for allowing this to go on. Because of him, and other inner workings of that dealership, I only lasted four months. One day, I didn't show up for work at the expected time. That is not how I operate, but in this case, I just couldn't be a part of how things were running. That day, I received a call around 9:30 a.m. from the grumpy service manager. He asked if I was coming to work and I said no. "I will not be working for you again." When he asked why, I was honest. I said that it was because of *him*. He was shocked. "Really, what do you mean?" That wasn't a question he should have asked, but he did. I let him know what I'd observed for four months. The way he treated customers, played favorites in the shop, had no systems in place, and his overall poor attitude. He didn't stay on the phone for long after that and ended with, "Sorry you feel that way." Things never changed at that place, and about a year later, they ended up closing for good.

I worked at another Chrysler dealership for about four years, which gave me a great foundation in the repair industry. Automobiles were changing drastically at the time with front-wheel drive, fuel injection, and electronic transmissions all becoming more common. I became a foreman there in a short amount of time and every problem I encountered was a challenge. Chrysler Corporation put me through a lot of factory training, and I used every bit of the knowledge I acquired to understand the new technologies. Transmissions were failing on a regular basis at the time, and we took care of our fair share. There were only a couple of us in the shop who were authorized to work on transmissions. I remember taking care of eight of them myself in a single week. Most were complete overhauls with a couple replacements mixed in. That has to be some kind of record. At the dealership, we were a close-knit bunch with regular get-togethers outside of work. This made for a pleasant working environment which we all cherished. The service manager was a real character. On phone calls that weren't going well, he would put the phone in the desk drawer and rapidly open and close it. "Oh, we must have a bad connection," he would say and hang up. This was his routine until one day when he was confronted face-to-face. I was on a road test at the time, but I saw that he got punched squarely in the nose. Another technician saw this happen and jumped to his aid, wrestling the customer to the ground while rolling down a hill. That was the talk for some time.

I also have some sad memories from that shop. One morning, I came to work and found out that a co-worker and close friend had gone home the night before, laid down as he was not feeling well, and never woke up. He was a great guy, and I guess the good die young. On another sad note, we returned from a Labor Day weekend break to find out the owner's son, who also worked with us, had fallen out of a tree while trimming branches. He is still paralyzed from the waist down. He had the character and resilience to pick himself up, dust off, and compete in wheelchair races across New England. He didn't wallow in self-pity; he kept positive and moved forward. His father, however, was devastated and was never himself again after that day.

One more tragic incident occurred at another Chrysler dealership where I had worked. The service manager, who was a very hard worker, was unlocking the door in the pre-dawn hours and collapsed from heart failure. He was only 42. I am telling these stories to allow you to understand that this is tough work. It's not the kind of job where you can get out of bed every morning, go to work, and come home and sit without some kind of regular exercise routine. Lower back and knee problems are common. You have to stand for long periods of time on cement floors, lifting heavy objects, working overhead or while bent over, twisting and pulling. All of these factors and more contribute to health problems. Walk into any garage and take a look at the older gentlemen working there. They move slowly or possibly limp. If they have been in the field for any length of time, they will be broken somehow. I'm not trying to discourage anyone from entering the trade; that is not my intention. I am merely telling these stories, and warning of possible hardships that can be avoided with regular strength and conditioning exercises. I have stayed fairly healthy in my nearly forty years in the field. I believe that is due to taking care of myself and using an inversion table (hanging upside down) for back relief. I highly recommend weekly and sometimes daily stretching on the inversion table. My martial arts training with its flexibility and stretching techniques has also been instrumental in keeping me healthy. I've had pulled muscles and strains as well as deep lacerations that have sidelined me for a day or two here and there, but that's about it. Now that I'm approaching sixty and have been teaching for a couple of years, I can feel the effects of my many years in the field but I am still not suffering as much as others who have fixed automobiles as long as I have.

Answering an ad in the local paper, I went in for an interview at another Chrysler, Jeep, and Dodge dealership. An interview for an automotive technician is a fairly easy process. I have always dressed nicely, but I feel that it is not a requirement. Proving that you know how to fix vehicles and that you are a team player will usually get you the job. If there are multiple applicants, then certifications, attitude, and references will set you apart. I have learned to never burn bridges especially if you are moving from one dealer to the next. The managers usually talk to each other on a regular basis at local meetings. They know that technicians are the backbone of the service department. I have been both a dispatcher and a service writer at various places and know that having a strong team of qualified technicians makes the job so much easier. It is difficult to call a customer and have to cover up for a technician that diagnosed wrong, broke something, or was incomplete in their original estimate. Things don't run so smoothly, especially if you have a shop full of incompetent people. On the other hand, when you can rely on the expertise of professionals in your shop, calling a customer with confidence in the repair makes for an all-around good experience for the customer as well as the staff. It is also the reason customers return again and again. Many times they will tell others about their experience which will grow your client base.

Service managers learn which technicians are in what shops. They keep an eye on the good ones and grab them if the opportunity arises. All technicians and service managers know that there isn't a tremendous amount of loyalty in the automotive field to commit to one location. Although this has gotten a bit better, money still motivates individuals to stay or go. For me, and most others, making an honest day's pay for an honest day's work is the bottom line. We just want to be appreciated. Working a forty-hour week and not receiving at least forty hours of pay, due to either a flat rate or some other pay structure, is a slap in the face and a sure way to create unmotivated technicians. Unfortunately, this happens everywhere every day. Some are constantly looking at others in the shop and comparing the type of work or hours of pay accumulated. Often, younger technicians will make more hours than others with more experience. For me, I always avoided the daily fluctuations and concentrated on the year-to-date numbers. That's what matters, and if it was not where I needed it to be, I would say something to management—not gossip with others and pout all day. This attitude is poison in a shop, and I have seen it in many places I have worked. My response is simple to the grumblers: "The door is right over there and that's why they put wheels on your toolbox." My only hope is that they will see the negativity and stress they bring to their lives, which can spill into other areas and can affect their health as well.

I answered an ad from the local paper for an automotive technician at a start-up Chrysler dealership. I was his first technician and watched the dealership grow around me. The dealership grew rapidly and filled up with a very diverse group of characters in the service department. There was an old-timer we called Buddha, due to his obvious resemblance to his namesake. He came with the building—Buddha had worked for the previous owners and stayed on. There was the loud joker who was always the center of attention. He was really fun to work with, creating or finding humor in everything around him. Then there was the guy I called the "quiet" character. He'd tell me stories of his life with his wife that would make me cringe. He would go on about how they threw things at each other, like lamps and vases. He made it seem like it was a normal relationship. I had no idea what to make of it. At work, if someone in the shop got on his bad side, he would wait until they pulled a car out and then urinate on the floor in their bay where they worked. This happened several times. If a customer aggravated him, he would climb onto their fender with one knee and fill up their windshield washer bottle with, you guessed it, urine. His nickname was Barney Rubble due to his resemblance to the Flintstones character. There are other stories from that dealership, but you get the picture.

The funny thing about that dealership was the fact that the family wanted nothing to do with the repair side of the business. I actually overheard the owner say that if he could cut the service department completely out of the picture, he would. That does not make sense to me, because without service, they will not have repeat business. You must keep customers coming back to your dealership for their automotive needs. When their vehicle gets old enough so that repairs are getting too costly, they can simply trade it in and start all over again. If they are treated right, they will have their friends and family come to you as well. Sounds like a simple formula for success that they simply did not understand. I believe that there is no reason why dealerships, or any automotive repair establishments, cannot retain at least sixty to seventy percent of their customer base by doing the right thing, honestly. With the millions of vehicles on the road, there is certainly enough money to be made the right way without sacrificing integrity. That dealer is still in business today, so something is going right for them.

CHAPTER 3:

"MY MECHANIC"

At every dealer I have ever worked, I would hear the customer bring up the term "my mechanic." If the customer needed some work and I let them know about it, I would hear thanks, but I'll just bring it to "my mechanic." "My mechanic" says I need this or that is another popular statement. When they hear their mechanic say they need something, there's no convincing them otherwise—what he says is gospel. You cannot change the customer's mind even if "my mechanic" is wrong. I often wondered what it takes to be the "my mechanic" in people's lives. I have had this conversation with several service managers, and many haven't seemed interested in trying to become the "mechanic" in people's lives—someone they can depend on and fully trust. Most just wanted to continue the day-to-day as it has always been by bringing in whatever comes to them by default. I believe that the criteria for being "my mechanic" should be something like this: a face-to-face relationship with the technician working on the vehicle, an accurate estimate of work performed with no surprises, finished in a reasonable time frame, completed at the stated time, a clear and confident explanation of the repairs made, an itemized bill with no hidden costs, a reasonable price for what was done, and a personable handshake and smile using their name when finished. Is this unreasonable to ask of any automotive repair establishment? I don't think so. It should be standard procedure, but unfortunately some of those steps are overlooked or not aligned with ethical standards in a lot of places. This raises concerns for me as an upstanding technician that wants the industry to do the right thing by our consumers. I don't think that technicians neglect their customers on purpose, I just think that they don't care enough. We know that people depend on their vehicles and therefore on their technicians. Doesn't it make sense to do everything they can to please the customers in order to keep them coming back? Perhaps they'd even tell a few people about their amazing experience, which would attract new customers. It's not rocket science, as they say!

Another thought I have approached management with is the idea of setting up dealers like doctors' offices or hospitals. I see it going something like this: the consumer picks a PCT, or primary care technician, which would be similar to a PCP, or primary care physician. The customer picks a PCT based on their experience and training report. They make appointments with their PCT based on his or her availability. If the PCT is not available in time, then the customer could see a TA, or technician's assistant, who is monitored by their PCT. I believe the best technicians would always be busy and this would motivate others to step up their training and attitude toward the consumers. Having such a structure would prevent customers from leaving a dealership or other repair facility if they have a bad experience. They could simply pick another PCT. The "my mechanic" now becomes their PCT. The service writers should be assigned PCTs to work with, and the PCT should be available to speak with their clients when necessary. In my experience, technicians don't like to leave the comfort of their work areas to speak with customers. In order to better serve their clients' needs they would need to be more involved personally. A vehicle is probably the second largest investment most people will make in their lifetime after owning a house, and it needs to be treated as such. Like I have stated already, many technicians only see a vehicle coming to their shop as a dollar sign and nothing more. Perhaps the PCT approach could change that. I would like to see a biographical write-up of all the technicians within a dealership posted in their waiting room. This write-up would list their schooling, experience, longevity, and philosophy, and include a personal note about their interests and hobbies. I believe that putting a face to their vehicle's technician will go a long way in retaining customers instead of a vehicle just disappearing into the garage and given to anyone for a repair.

These days, when the consumer is faced with many choices in auto repair, reaching the consumer on a more personal level is critical. Using technology, educating the consumer, and meeting their personal needs are more important than ever. I have listed only a few of my suggestions, but there are many more ways in which shops can continue to keep customers returning for years to come.

CHAPTER 4:

MECHANIC VS. TECHNICIAN

Before I continue with my personal narrative, I am pausing to explain the difference between a "mechanic" and a "technician." What comes to mind when you hear the word *technician*? Do you picture someone who looks neat with clean hands, precise measuring equipment, and a wealth of knowledge? That's the image that the industry had in mind when they introduced the word *technician* to the automotive industry.

How about the word *mechanic*? Most people think of greasy hands, dirty clothes, and a sub-par education. The industry was looking to change the perception of those who repaired vehicles for a living. With the introduction of electronics in automobiles, bringing your car or truck to a mechanic just didn't seem to fit. Bringing it to a technician, now that sounded like a great choice. It's amazing to me how just one word can change our perception. Throughout my career, I've heard people say, "I used to be able to do a lot of my own work, and now I open the hood and don't even know where to put the oil in." I guess that makes this trade a whole lot more respected. Changing the name to *technician* is only natural.

Both mechanics and technicians exist in the trade today. I present this dichotomy to my students so they can understand the difference and choose their own path. I draw two columns on the board with the heading of mechanic on the left and technician on the right. In both columns, there are A, B, and C levels, which I place in descending order in the center of the columns. Everyone starts out at a C level. At that stage, they are starting to get involved in a lot of small repairs, such as tires, oil changes and filters, but they can choose which column to start climbing, the mechanic or the technician. Are they organized and take pride in their work, or are they sloppy and incomplete in the repair? Do they leave the customer's car clean and free of handprints, or is the seat moved and the radio stations changed? Do they double-check their work? Answering these questions will put them in one of the two categories that should not need further clarification. They will remain C level for roughly two years, and then they would be moving into the B level mechanic/technician. They begin to gain confidence in their ability to repair vehicles with less supervision. Within the B level, they take on more difficult work and increase their diagnostic capabilities. Most who choose to remain in the mechanic category may not even know that there's a difference between a mechanic and a technician, but those who choose to be technicians surely understand the difference. They take pride in their work and accept that pleasing the customer is their number one priority. Their business is built on customer retention and loyalty. They become the personalized *technician* customers trust. Mechanics, on the other hand, may not be repairing the automobile with tomorrow in mind. They are working for today's pay and can't see past that. Mechanics don't often take the time to diagnose vehicles properly. Have you ever heard the term "parts replacer"? This refers to a mechanic that replaces parts until they get it right. If you understand this, then you will understand that mechanics may cost the consumer more money than a technician that may charge a little more per hour.

After a technician has been in the trade for around six to ten years, depending on the individual, they may move into being an A-tech. I said *may* because some remain a B-tech for longer or may never move onto being an A-tech, depending on the individual's work ethic. They just don't have the ability or skill set that it takes. An A-tech can work on any automotive system with confidence and spend whatever time is necessary to complete the task. They are the go-to guy in any shop. If they don't have the answer, then they know where to find it. A-techs make sure they keep up with the latest technology and products to help them remain on the cutting edge. I remember eating lunch in a private meeting room watching the latest training video at the dealership when I was the foreman. My service manager walked by the door and spotted me. He stopped, backed up, and looked in. "That's exactly why you are the best at what you do," he said. That was a good feeling—one you could only get by being "all in." There can be A-mechanics who know what they are doing but are sloppy and don't work hard to better themselves like a technician does. Moving forward in this book I will refer to anyone repairing a vehicle with ethics and credibility as a *technician* in order to show respect for all that we do. I will use the word *mechanic* in order to bring attention to a judgement call I am making on their morals and stature out of respect for the trade.

There is an accreditation organization named ASE that provides credentials to automotive technicians. Through a series of lengthy tests, technicians will be provided with supportive documentation that allow them to become masters of their craft. Advanced levels are also available to technicians who want to go a step further. ASE serves all the transportation clusters, such as cars, trucks, busses, collision repair, and machinists. Upon passing these tests, technicians earn certificates of achievement, which they can proudly display for their customers. Because the automotive industry changes so rapidly, technicians need to recertify every five years with another round of testing in any area that expires. This is one way for the consumer to find a repair facility that is doing the right thing. Not a lot of mechanics bother to get certified through ASE, and I have heard some say that it is a waste of their time. This might be true if you are not interested in bettering your position in a company or honing your craft. ASE certifies to the consumer that you care enough to go above and beyond others and prove you are someone they can trust. ASE credentials are displayed with pride at any location that employs their skills. In my school, I have students test for what is known as student certification. The same testing applies, but fewer correct questions are required to pass and the two-year field experience is waived. The student certification is good for two years. After students pass these tests, I can see changes in their attitudes about the profession. They seem to listen a little more attentively, take better notes, and horse around less often, and they just show an all-around better effort. This effort carries over to their job placement. Employers can see the credentials that the students have earned, and in some cases, they receive a higher starting wage.

This may be a good time to tell you about an incident with a coworker, "Carl", a few years ago when I was still working in the field. Carl brought in a vehicle to look at the brake noise that a customer was complaining about. After looking over the vehicle, Carl brought the list of needed items to the service writer, who would talk to the customer. The needed repairs included rear brake shoes, wheel cylinders, resurfaced brake drums, and new tires. Carl received the okay from the customer to complete the needed repairs to the brakes, but they did not want the tires replaced. An hour or two later, the vehicle repair was complete. The customer paid the bill and was on his way. It just so happens that our shop was located next to a tire store, which was the customer's next stop to get the recommended replacement of the tires. Shortly after the customer left our shop, we received a phone call from the tire store. It seems that the tire store was recommending that the customer have the rear brakes replaced. Of course the customer produced the receipt for the brakes that were supposedly just replaced at our shop. A dilemma began to take shape. The service manager and I, the foreman, took a stroll next door to the tire shop to investigate the allegation. Sure enough, the brakes were not new, and the wheel cylinders were leaking. It seems that none of the repairs that the customer paid for were performed. That did not sit well with me, the customer, or my manager. My manager and I went back and questioned the *mechanic* who worked on the car. I searched the barrel, and under his work bench I found the brand new parts that were never installed on the customer's car. Shame on him! I was appalled to have him in my shop. I had a meeting with the owner of the dealership, the general manager, and the service manager. There was a discussion as to the proper course of action for this atrocity. There was no question in my mind—get rid of him ASAP. Turns out, they believed that he would learn from this and be a better *mechanic* after three days at home without pay. That was the way he was punished, a slap on the wrist and don't do it again. I worked with him for many years after that and kept a close eye on him. Can I say for certain that something like that ever happened again with him? I cannot. This was just one incident, but things like this do take place every day in the automotive industry.

After I began teaching, I was an outsider looking in and became a customer of the repair industry. I sent my son to a local repair shop for an inspection sticker after I had performed many repairs on his vehicle at home. I was astounded when he returned from that shop with a list of needed repairs including tires, ball joints, and an oil leak. I promptly called the shop and asked if I could come down with the vehicle and have them walk me through the needed repairs. At that point, I had not revealed that I had any knowledge about cars. They grabbed the mechanic who'd rejected my son's car so we could discuss the needed repairs. First, he pointed to the tires: "Take a look at those tires; they're all worn out." I asked him what the tread depth was and he said that he could tell that they wouldn't pass inspection by looking at them. At my insistence, he measured the tread depth and they were above the minimum thickness of two thirty seconds of an inch.

"Okay," he said, "I'll pass the tires, but the other items still need attention."

"Show me," I said. He raised the vehicle in the air and proceeded to show me a "bad oil leak." I explained that there was no oil leak as I'd repaired it prior to bringing it to them for inspection, but there may have been some old oil remaining on the undercarriage.

"All right, we will pass that, too," he said. By now, they were beginning to realize that maybe I knew a little more about automobiles than I was letting on. One of the last items on the list was ball joints, which I had also just replaced. He placed a jack under the vehicle and showed me movement in the wheel assembly coming from the ball joints. He said that the movement was unacceptable and would result in a failed inspection. I asked him what the measurement was for the amount of play. He kept saying, "Look how much that moves—that won't pass." Now the game was up, and I had to reveal my identity because I wanted the exact measurements. I had looked up the specifications for the maximum amount of movement allowed. The state law stated that the ball joint measuring tool needs to be located in the state inspection bay, so I asked him one more time for the measurement. Now the excuses started to fly. "Joe is out today and he has it locked up in his tool box." As I was about to leave, the manager grabbed me and told me to send my son back with the car the next day and he would pass it. That would have cost the average consumer upward of a thousand dollars. It allowed me to see the consumer side of auto repair—which I did not like.

All mechanics and technicians at one point or another cut a corner or two when making a repair. They may not spend the required time to change the coolant properly or replace the one spark plug that is almost impossible to access. Sometimes the whole dealership is involved in dishonesty without even knowing it. One dealership I worked at had a big, elaborate-looking machine that was nothing more than a fluid vacuum that they used to service automatic transmissions. You would shove the wand down the dipstick tube, suck out whatever fluid it could reach, and refill the transmission with the same amount of new fluid, usually about three quarts. Most automatic transmissions hold a total of ten to fourteen quarts. Does that sound like a proper service? It is not, and the customer is charged for a full service. To properly service an automatic transmission, the transmission pan should come off and be inspected for any debris that may indicate future transmission problems. The filter should be replaced and the pan properly resealed. Another method would be to completely flush all the old fluid from the transmission and replace it with new fluid using a transmission flushing machine.

That same dealership would change brake fluid by using a turkey baster to suck old fluid from the master cylinder and then refill it with new fluid. That leaves old fluid throughout the system that should have been flushed out. The customer is again charged for a full service. When everyone in the service department, including the manager, agrees to these methods it is hard to change their ways.

Whether a dealership or an independent repair facility, technicians will usually find an easier way to perform a repair. This does not necessarily include dishonesty; they have just performed the repair so often that they have figured things out. One part of the repair procedure has been modified or eliminated by savvy techs. Often, a particular shortcut will travel around the shop and migrate into the surrounding shops and become commonplace for that type of repair. If time can be saved, which equates to money in the eyes of repair professionals, then this repair will be done using the new technique every time. I have originated these new techniques in the past and also incorporated someone else's repair procedure into my routine as long as quality does not suffer from it. Sometimes these new procedures make it all the way to corporate, and a new repair procedure will be released to the industry.

There are certainly a lot of positives to using dealerships for auto repair. One of the key factors to the success of the dealer service department is the technicians they employ. They require extensive training to bring them to the master level with the product they support and then regular training to keep them current. Most dealerships will have several, if not all, A-techs that are masters for each manufacturer product that they sell within the dealership. In order for the dealership to get paid on most warranty claims that they submit, the technician working on a particular job will need to be certified in that specialty area. That's why A-techs that are master certified can repair all aspects of the automobile without worrying if the repair they are performing is covered under their specialty area. Another positive to using dealerships for auto repair is that they almost always use original equipment manufacturer (OEM) parts in all of the repairs they make. Which, in my experience, fit and perform better and outlast the aftermarket parts. I say that dealers *almost always* use factory parts because there are times that a part may not be readily available from the manufacturer, or it just may be too expensive to use a factory part in an older vehicle. The customer is usually made aware of this and given the opportunity to give it their okay. The dealership will still honor any warranty on the repair for labor, but the part warranty may be a little shorter.

CHAPTER 5:

TECHNOLOGY

There are times when a vehicle has a problem that is very difficult to diagnose. If this occurs at an independent shop, they may bring it to the dealer to be looked at or diagnosed if they feel they have done all they could at that level. A perfect example of this is when Sears had a thriving automotive repair business in the Boston area. Vehicles would come into my dealership with a starting or charging system problem after an attempted repair at Sears. They always had a new battery, new starter, and a new alternator installed. When I would get the vehicle, sometimes it would be a simple repair such as a loose wire or a bad fuse. There were times when it became a little more complicated, but the customer had already paid Sears. Sears made their money and then sent the customer on their way to have it diagnosed properly. When a dealer technician has a difficult problem, we are the last line of defense. They will run it by a fellow technician or a foreman if the shop has one, after spending time on their own. If the problem proves to be even more than the shop can handle, they will call a factory hotline. In my experience, the hotline looks at most of the same material that a smart technician will have already looked at such as technical service bulletins (TSB's) and online information. If the problem still can't be resolved through these methods, the hotline has communication lines open to the engineers responsible for the product support. Technical advisors are available to travel to regional dealerships to help with the diagnosis of these difficult problems. The tech advisor has direct access to engineers and all the latest information and equipment necessary to address whatever problems cannot be addressed with conventional methods. I recall one day walking up to a tech advisor that I had known for many years. He was working on a problem vehicle. He had the door open, a phone on each ear, a laptop in front of him, a factory scan tool plugged into the diagnostic connector of the vehicle, and assorted wires coming out from under the hood attached to a multimeter streaming live data. He looked at me and asked, "Can you believe how far vehicles have come?" I chuckled and agreed because we both began our careers in the age of carburetion and much simpler systems. I shook my head and went about my business with a smile on my face. This *is* how far we have come. When I started in this trade, we had two wire meters and a timing light for diagnostic equipment. Now we are accessing software using sophisticated scan tools and in many cases reprogramming that software in the field.

It is true that electronics have completely taken over the automotive world. When I began in this industry in the late 70s, most everything was mechanical. Electronics were found only in the ignition, lighting, and instrumentation systems. Engine vacuum was the power source for actuators used for moving valves and heating doors. Some windshield wiper and emergency brake systems also used vacuum for movement. With other systems, engine temperature changes moved valves that switched the vacuum from one place to another as dictated by temperature. Carburetors were common, and many technicians (myself included) got very good at diagnosing and rebuilding them. Although all of these systems worked as well as could be expected, the accuracy of the devices was limited. Thus began an electronic revolution in automotive technology that continues to this day. Many of the changes were necessary in order to keep up with federal emissions standards.

Just about everything that you see under the hood of an automobile is needed for this: injecting fuel and bringing air into the engine in order to move a piston that turns a crankshaft. Fuel injectors can be controlled so precisely that no matter what the operating conditions are, the exact fuel-to-air ratio is maintained. Sensors are in place to monitor weather conditions, air temperature, altitude, vehicle speed, and driver inputs such as gas and brake pedals. Although the basic engine design has not changed all that much, the ways that fuel and air are brought into an engine has. Precise opening and closing of the valves is now computer controlled throughout the industry. Electronic-controlled solenoids channel oil pressure to move mechanical parts. This allows engineers to program more precise management of the intake and exhaust valve movement.

Electronics have made their way into just about every system in the vehicle. Brakes began as a simple stick that pressed a block of wood against a wheel at the turn-of-the-century. They have evolved into drum and disk brake systems still in use today. Power brakes, anti-lock brakes, traction control, and vehicle stability control are some of the features that have been added throughout the last few decades. Today's brake systems incorporate computers that can slow a vehicle down when approaching an object without driver input and steer its way out of a skid using the electronic features of the brake system. Steering systems have also gone the way of electronics to the point that the driver does not need to have their hands on the wheel to get to their destination. A vehicle has already been developed by General Motors without a steering wheel or driver pedals. This technology will bring a new set of rules lawmakers need to consider, such as the perils of drinking and "driving" and who is at fault if a collision should occur.

Customer conveniences have been added, such as heated and cooled seating, radios with integrated navigation systems and Bluetooth technology, electric remote starters, automatic temperature control with zones for each passenger, heated and cooled cup holders, and a long list of other features. Passenger safety has not only become government mandated, it is used as a selling feature by manufacturers. Crash tests that are rated high in an impact will sell automobiles. I recently pulled up to a vehicle late at night that had run off the road and into a tree. The driver was unscathed because all the airbags had gone off. If this had taken place just fifteen years ago, I would have pulled up to a completely different scene.

In the sixties and early seventies, the "need for speed" took hold. These decades became known as the muscle car era! There was a horsepower war between the big three—GM, Chrysler, and Ford. Some—like the Mustang, GTO, and Charger—were huge successes. Others like Ford's Edsel were not. Bigger engines meant larger openings in the intake systems. Manufacturers found ways to *force* more air and fuel into the engine using turbochargers and superchargers. The power of these muscle cars created a dilemma—air pollution. This was at a time when scientists were just beginning to understand the effects of industry and automobile pollution on the atmosphere. On December 2, 1970, the EPA was formed. That agency has set and implemented emissions standards to control pollution on passenger vehicles, heavy duty trucks and buses, construction and farm equipment, locomotive and marine engines, and even lawn and garden equipment. The automotive engineers search for solutions to this problem, which is ongoing today. Fuel injection, catalytic converters, EGR valves, AIR pumps, and evaporative systems, to name a few, have been added to engines to control the ongoing pollution crisis. Manufacturers have even gone so far as to add start/stop technology to automobiles where the engines will shut off when the vehicle comes to a stop and restarts seamlessly upon removing the driver's foot from the brake pedal. New passenger vehicles are 98–99% cleaner for most tailpipe pollutants compared to the 1960s, but this is still not enough, given the millions of cars and trucks on the road. These government agencies have driven a lot of the technological advances that we see today. In order to control emissions, precise control of all systems that we already spoke about needed to take place.

The generals in the armed forces are another driving force for new technology. Navigation, radar technology, and battery technology are just some of the forward-thinking elements that have come from the military contracts. They will have a need in a certain area and then challenge the industry to come up with the best solution to the dilemma. This is how innovation takes place even today.

As an automotive technician, I have adapted when needed to accept the change and challenged myself to be at the forefront of repairing this technology. That is what technicians need to do in any industry. Humankind does not settle for what's out there; we constantly set our sights on improving existing products or completely change the way we look at it.

So what's next? Gasoline-electric hybrid technology has been around since the early 1900s and has made a comeback in the last decade or two. This is more of a link to fully electric vehicles than it is a mainstay. Many believe that hydrogen power is the holy grail of the future due to its *water in/water out* theory of operation. The process of adding water to a car, separating the hydrogen from the oxygen, burning the hydrogen in the engine, and reforming water after combustion, is an exciting possibility. Tesla has made great strides in fully electric transportation, but battery technology and recharging infrastructure has somewhat limited its potential. *Automotive News* estimates that 20% of all automobiles on the road in 2030 will be electric. Another estimate is that the earth's ozone will be fully restored by 2065 if we continue our current pace of downsizing the amount of internal combustion engines. I wonder what effect this will have on the large oil corporations and the unrest in the Middle East.

CHAPTER 6:

SELF-EMPLOYED

The ultimate goal of workers in many professions is to own their business. Automotive is no different. After ten years, I decided to try it. I attended franchise shows in hopes of finding something in the automotive field that would suit me well. There was a tune-up franchise that did interest me, but the expense was a little higher than I thought it would be. The reason I wanted a franchise was the support I would receive from the corporation. I was young, so being self-employed was a little scary. This automotive tune-up franchise would be by my side to help me find the right location, set up the building, stock the right supplies, train me and my staff for success, and offer any ongoing support I may need. After a week or two of dead ends on raising enough money to get started, I dropped the idea of a franchise.

In asking around, I found a gas station that was looking to lease the two bays attached to their building for automotive repairs. I thought this would be a great way to get started, so I went down and talked to the owner of the station. The owner, John, was a decent guy who inherited the business from his father, which included another gas station directly across the parkway from the one I was looking to lease. The location could not have been better—right on a busy parkway about a half mile from where I was living at the time. John did not want to get involved in automotive repairs, but he wanted to profit from owning the property. His second location already had someone working out of the service side of the business. John and I reached an agreement that included a monthly rental charge and a small percentage of my monthly income. I had two bays with lifts to work out of, and the shop already had most of the equipment I would need to get started. With the help of a friend, I built storage cabinets and benches and organized things the way I wanted it. I had a *Charlie's Automotive* sign made and proudly displayed it at the street right under the big Mobil gas sign. I was now open for business.

I grew up not far from this location, so I had lots of friends and family to provide me with the initial wave of work that I needed to get started. John was diligently selling work for me at the pumps to ensure that his monthly percentage was adequate. I was set up for success. I worked alone for about six months or so, and things went pretty smoothly. For the most part, John and I worked well together. I say *for the most part* because there were times when he did not honor the separation of my business and his. John began to cross the line in thinking that I worked for him. We straightened that out on more than one occasion. I also hired a lawyer and set up my business as a chapter S corporation in order to protect my personal life should something go wrong.

As time went on, I could not handle the amount of work that was rolling in, so I hired a friend to help out. I set up my new employee with all the taxes and insurances that were necessary. In doing so, I realized a business has to pay almost the same amount in taxes and insurance that they pay the employee. A small business like I had with one employee is difficult when state and federal have their hands out looking for their cut. I tried to get an explanation for an extra five percent that I was charged in state taxes for having a small business. I made several attempts with different departments within the state to get an explanation for this. What I was told, and I still cannot believe, was that it was "for the privilege of doing business in Massachusetts." That was a tough pill to swallow. Health insurance was also expensive to carry for myself without belonging to a group plan, and adding an employee further strained the budget.

I gave myself one year from the time I set up my business to see if it was worth it to be self-employed. What I found out was that having your own business is stressful, and it's always on your mind. I remember waking up in the middle of the night on several occasions stressing about something I needed to do or something I forgot. Certain customers stay in your head as well. I had a couple of customers who refused to pay and picked up their cars with a spare key after hours. I have had police involved on a few occasions and have taken others to court only to have the customer not show up. I would be unable to collect for the outstanding bill because the customer could refuse everything that they were asked to do by the courts and walk away. The small claims process is not very effective, and it cost me time and money. I believe I had a fifty percent collection rate on six customers.

Of course being self-employed does have its advantages. Coming and going when it is necessary without having to answer to anyone is one benefit. Declaring what holidays I will take off and what times my business will be open is another. But you must remember that there are no paid holidays. If you take time off, the income stops, unless you have several employees you can count on. Even then, you are obligated to pay them for the holidays. It seems someone always has a hand out.

After the year, I gave it up. I found that I could make more money working for someone else and punch out at the end of the day without any worries. I went on to sleep better and continue a successful career elsewhere.

CHAPTER 7:

FIRE AND SAFETY

Safety is important in the automotive industry. We need to protect ourselves from the fluids and chemicals we use daily—some of which are very harmful. Something as routine as changing oil can have long-lasting effects. Material Safety Data (MSD) is required to be on hand anywhere that stores or uses that particular fluid. The MSD provides all the information needed when dealing with a chemical, such as fire handling, storage, spill cleanup, and first aid, should the chemical be ingested or inhaled. Below is a section from the MSD for motor oil and used motor oil. As you can see by this one item, technicians are at risk every day and should wear proper respiratory equipment and gloves and coverings of exposed skin when coming into contact with the more harmful substances. While this is not commonplace in most shops for chemicals, technicians know when to take these steps—they just choose not to.

Below was taken directly from Massachusetts right to know handbook:

DESCRIPTION OF NECESSARY MEASURES WHEN DEALING WITH MOTOR OIL

Inhalation IF INHALED: Remove person to fresh air and keep comfortable for breathing. Call a POISON CENTER or doctor/physician if you feel unwell. If breathing is difficult, oxygen should be administered by qualified personnel.

Skin IF ON SKIN: Wash with plenty of soap and water. If skin irritation or rash occurs: Get medical advice/attention.

Eyes IF IN EYES: Rinse cautiously with water for several minutes. Remove contact lenses, if present and easy to do. Continue rinsing. If eye irritation persists: Get medical advice/attention.

Ingestion IF SWALLOWED: Do NOT induce vomiting. Immediately get medical attention. Call 1-800-XXX-XXXX for additional information. If spontaneous vomiting occurs, keep head below hips to avoid breathing the product into the lungs. Never give anything by mouth to an unconscious person.

Most Important Symptoms/Effects

Acute Harmful if swallowed. Causes skin irritation and eye irritation. May cause allergic skin reaction, asthma, allergic reactions, respiratory tract irritation, and central nervous system depression. Causes damage to kidneys, central nervous system, and lungs.

Delayed May damage fertility or the unborn child. May cause cancer and mutagenic effects. Indication of Immediate Medical Attention and Special Treatment Needed. If needed, treat symptomatically and supportively. Treatment may vary with condition of victim and specifics of incident. Call 1-800-XXX-XXXX for additional information.

Precautions for safe handling Keep away from flames and hot surfaces. Wash thoroughly after handling. Use good personal hygiene practices and wear appropriate personal protective equipment (see section 8). Spills will produce very slippery surfaces. Used motor oils have been shown to cause skin cancer in mice after repeated application to the skin without washing. Brief or intermittent skin contact with used motor oil is not expected to cause harm if the oil is thoroughly removed by washing with soap and water. Do not enter confined spaces such as tanks or pits without following proper entry procedures.

Do not wear contaminated clothing or shoes.

Conditions for safe storage: Keep container(s) tightly closed and properly labeled. Use and store this material in a cool, dry, well-ventilated area away from heat and all sources of ignition. Store only in approved containers. Keep away from any incompatible material (see Section 10). Protect container(s) against physical damage.

USED MOTOR OIL EMERGENCY OVERVIEW:

APPEARANCE Liquid, black, and viscous (thick), petroleum odor.

WARNING! PHYSICAL HAZARDS Combustible liquid.

HEALTH HAZARDS May be harmful if inhaled. May be harmful if absorbed through skin. May be harmful or fatal if swallowed. May irritate the respiratory tract (nose, throat, and lungs), eyes, and skin. Suspect cancer hazard. Contains material which can cause cancer. Risk of cancer depends on duration and level of exposure. Contains material which can cause birth defects. Contains material which can cause central nervous system damage.

ENVIRONMENTAL HAZARDS Product may be toxic to fish, plants, wildlife, and/or domestic animals.

In addition to chemical hazards, there are also hot or moving parts that can cause irreparable harm, such as burns, deep lacerations, or in extreme cases, loss of fingers or limbs. You will walk into lift arms and undercarriage components from time to time, causing cuts and lumps on your head that will aggravate you. It comes with the territory, as they say. Debris in the eyes is an ongoing issue, which is easily prevented by wearing safety glasses that most in the industry do not utilize. In my school, I stress the importance of safety glasses and make it a requirement for all students working in the shop. I must say that overlooking the common sense approach to safety is the number-one factor in bodily injury.

Sometimes, people just do not think something through before executing. One dealer I worked at *almost* had one of the most tragic accidents that could have been completely avoided with a little common sense. I looked across the shop in time to see one of our mechanics removing a fuel pump from a gas tank that was apparently still full of gas. What I saw astounded me. The mechanic was completely covered in gas dripping down his arms and covering his pants. He had a lit cigarette hanging out of his mouth. Gas was also puddled up around him running down the floor as he was trying to hold the tank up with one hand. With the other hand he was trying to place the jack in a better position to hold the tank up. There was a light containing an incandescent light bulb hanging beside him, swinging wildly. As I was screaming across the shop, the drop light fell, and believe it or not, the bulb did not break. Several of us rushed to his aid and secured the ugly scene. If that drop lightbulb had broken when it hit the floor or if the cigarette dropped from his mouth, there would have been no rescuing him, and he would have perished tragically in a ball of fire while we called for help and tried to get close enough to extinguish the flames. I have no doubt that this event also would have touched off a chain of events that would have changed that dealership and me forever. Very lucky indeed!

We work around gasoline and hot engine parts all the time, which when not respected can lead to fires. Forgetting to tighten the fuel line, or not connecting it to the engine at all is one leading cause of fires in automotive shops. Technicians are constantly trying to work as fast as possible to increase production, and they sometimes overlook details. I have done this myself and started fires that are very scary to witness. Jumping into action and grabbing a fire extinguisher can limit the damage. Most of the time, other than those around you, no one ever knows that the fire took place. Cutting and heating parts using acetylene torches are another source of starting fires. Usually a fire started in this manner is quickly contained. I witnessed a fire started by a technician who was cutting with the torches close to the floorboards around the back seat area of an automobile. The car was up on the lift, and with all the windows closed, no one realized the interior of the car was in flames. By the time they brought the car down and put the fire out, the interior had extensive damage. I'm not sure how the car was repaired or replaced after that, but it would not have been a pleasant conversation with the owner of that particular automobile.

I have also seen the aftermath of vehicles falling off lifts. Several years ago, a technician had just stepped out from under the vehicle to get the parts he had ordered. He looked back after hearing a very loud crash. What he saw was the Isuzu Trooper on its side in the very spot where he had just been standing. Close call indeed! It had been raining that day and the lift arm locks were broken, which contributed to the arm of the lift slipping off the frame of the truck. Fortunately, the Trooper was a used car that belonged to the dealership. We dragged the truck out sideways using a fork truck and rolled it back onto its wheels with about eight of us pushing. I'm not sure if it was ever fixed or declared a total loss.

These are just a few of my experiences. It is scary to think about the deaths that have occurred from, in most cases, easily preventable accidents. The agencies that create workplace rules and regulations undoubtedly have saved many lives and should be commended for their work. They put these rules in place to prevent accidents, but in my opinion, they do a poor job of making sure they are adhered to. Spot checks should be performed much more often. Usually when these spot checks are performed, the shop has advanced notice, giving them time to get things back in order. There is a law in place for private industry that is called the "right to know" law. Under the Massachusetts Right to Know Law, communication of information to employees is accomplished in three ways:

Material Safety Data Sheets or MSDSs: These brief documents, obtained from manufacturers and suppliers of toxic and hazardous substances, are the primary sources of information under the law. They include information such as chemical identity, physical properties, health and safety hazards, safe handling procedures, and spill, leak, and disposal procedures.

Container Labels: The chemical contents of certain containers must appear on the label. Ideally, the label alerts workers and supervisors to the presence of toxic or hazardous chemicals. Also, such information is useful to an employee's physician and workplace health and safety committee.

Employee Training: Employers must provide annual training to employees on the hazards they work around and safe handling procedures of regulated substances in the workplace. New employees must also be trained within thirty days of employment.

Some employers bring in a private company to handle educating employees and labeling containers. They also work with employers to make sure their shop is OSHA compliant. In my experience, not all shops comply with this law, and that is a shame. Here is the link to the full Massachusetts employer handbook containing this law:

https://www.mass.gov/files/documents/2016/08/uf/ma-rtk-employer-manual.pdf

CHAPTER 8:

DEMOGRAPHICS

I have noticed how demographics plays a role in automotive repairs. It seems that when I have worked in a well-to-do area with a higher income per capita, people were a lot fussier and penny pinching when it came to having work done on their automobile. Selling a customer the required repairs took a little more time. They wanted to see exactly what I was talking about and explain to them why they needed it. They usually had alternate transportation so time was not an issue. I see no problem with this and believe that every motorist would be better off if they took this approach. Once you explained everything, they would then shop around for the best price. Service writers in these areas knew how to work with these customers explaining why they wanted us to do the repair rather than a less expensive shop. A lot of times the service writer would adjust the cost to make the sale. At shops within a less affluent area, customers that needed additional work on their vehicles would not haggle. More often than not, I would hear the customers say "Just fix it, I need my car back. And when will it be ready?" Their car is their life. No car, no employment—it's that simple. They couldn't afford to have it break down or be without it for any length of time. Think about losing your transportation for even one day. It does not put you in a good situation with the need to get to work or grocery store. Fortunately, there are alternatives that may get you by for a day or two such as Uber or Lyft or perhaps a rental car. You may even be able to catch a ride with a friend or coworker. A repair that takes longer than a couple days is going to be a problem for most lower or middle class families, and that's why maintaining a vehicle is so important. In my experience, letting even an oil change go too long can have catastrophic results that cost time and money. Money can be saved in other ways, too, such as receiving a ticket for having burned-out lights or an expired inspection sticker.

As a technician, I would always take the time to make sure that my clients had reliable transportation regardless of income. The single mother struggling at a waitress job and the COO of a multimillion-dollar company, still need to get from point A to point B without the hassle of a breakdown. All technicians in the automotive trade would do well to realize this and look over vehicles and report the findings. The customer can say no, but you have at least reported to them what you have found. A safety item needs to be addressed immediately. How would the technician and the establishment feel if a customer who has just left your shop gets into an accident due to a worn-out part that you failed to find and report. Perhaps there was no accident, but the customer had just spent hard-earned money and left with an unexpected indicator light on in the dashboard. Technicians need to work toward making a customer for life not a one-time visit. Take the time to top off fluids, check lights, and set tire pressures every time a vehicle comes to you for service in order to prevent that awkward conversation.

CHAPTER 9:

TOOLS

Tool purchase is an ongoing investment in the career of an automotive technician. Most invest upwards of twenty to thirty thousand dollars or more in tools over the span of their career. I made my first tool purchase from Snap-on tools working at that first gas station job. Tools were more affordable back then, and I would have made a larger purchase had I known how much the Snap-on price would increase. Snap-on tools are top of the line according to all automotive professionals. Anyone who has spent time repairing automobiles owns at least some Snap-on tools and has the receipts and depleted bank account to prove it. They are not cheap but are invaluable when it comes to counting on a quality tool. They are only available through vendors who visit repair shops on a weekly basis. Other tool vendors that visit shops weekly include

Mac Tools, Matco Tools, Cornwell Tools, and miscellaneous independent vendors that carry other names such as Stanley, S-K, or Blackhawk. Craftsman tools are also a good choice which were exclusively available at Sears stores. Sears has since gone out of business, and Craftsman tools are available at other home repair and hardware stores. All of these manufacturers provide quality tools, and I own tool sets from most of them. There is a lifetime warranty on tools from all of them, so making a purchase from any is a good investment.

When mechanics in any capacity see tools and equipment while they are out, we tend to gravitate toward them. This is something we cannot help; it's part of our DNA. We make our living with tools, and anything that makes the repair process easier grabs our attention. Flea markets and yard sales are great places to get deals on tools, and we always have our eyes open. When our partner suggests attending one of these events, we go, but have our own agenda in mind. Where are the tools? Snap-on created a sign that most technicians either post on their tool box or live by: "I make my living with Snap-on tools. Please don't ask to borrow them." The reason is that there is nothing more aggravating than needing a tool, and then having to go retrieve it from someone who borrowed it. This tool etiquette is a universal rule throughout the automotive industry and needs to be strictly adhered to or one may find out the hard way. The owner of the tool will let the borrower know if things are not returned promptly. That may be the last time they borrow from that person. When professionals work together for an extended period of time, borrowing a specific tool is not usually an issue because they both know the etiquette involved. They will use it, clean it, and return it right after use or certainly within a day. If you borrow a tool more than once, it needs to be purchased—plain and simple.

Technicians sometimes make tools for a particular repair. Anyone in the field for any length of time has a special place they keep the self-made tools. There could be ground-down sockets or wrenches, bent wrenches, shortened tools, or tools created completely from scratch. These tools are invaluable in certain situations. Automotive manufacturers do not always think about the repair industry when they assemble vehicles at the factory. Sometimes what should be the simplest repair turns into disassembling half the car to remove a part. Shaking your head and bitching about it while working on the repair seems to do the trick. This is where homemade tools are usually created. If time can be saved, we will find a way. There are also occasions when two parts will just not separate. I find that if I walk away, take a breather, and return, the items fall apart much easier. Bitching and whining won't usually help here, but it happens regardless and relieves some stress.

CHAPTER 10:

TEACHING AUTOMOTIVE

I began my teaching career by answering an ad for automotive technicians that want to teach but have no experience. The year was 2012, and I had 34 years under my belt in the automotive field. I had spent the latter part of those 34 years trying to change careers to do something different, but teaching was one of the fields that had not been on my radar. Fifteen years earlier, my old vocational instructor was trying to convince me to take his position at my alma mater as he was retiring. I was a foreman and making great money at the time, and I didn't feel confident enough to teach. As time went on, I was never fully invested in the automotive field where I was content doing what I was doing. But as they say, that was all I knew. I loved to cook, so at one point in 1993, I decided to go to culinary school to become a chef. Chefs were gaining respect and were regarded as "cuisine creators" and held in high regard. I wanted a piece of that energy. So 18 months later, I graduated night school and received a chef's certificate. Did I mention that I was unafraid to take risks? I quit my automotive job to start cooking. Well, it was no fun having to work for less money at such a fast pace it made my head spin. The fast pace starts from the minute you set foot in the door and sometimes even before that due to deliveries sitting outside. Things really heat up when the lunch or dinner crowd rolls in. At the places where I worked, the line cook was responsible for several stations at once. I was moving between the grill, the fryer, and the broiler like a madman. When the rush was over, it seemed like only about 15 minutes had gone by. There is no time to think. Reality was that usually one and a half to two and a half hours had gone by. When the rush subsides, you take a deep breath, look around. and say to yourself, *That was crazy.*

Most workers in the culinary field have a lot of respect for automotive technicians. The other cooks I worked with could not believe that I walked away from a career like that. Needless to say, my cooking career was short-lived, and I went back to the automotive field. The experience made me appreciate fixing automobiles—for a little while anyway. At another point in my career, I went back to college to work in the computer-aided design field which was another up-and-coming trade where technology was changing daily. I figured that I could design automotive parts or maybe even automobiles. Again, I never pursued that career path and dropped out of the program realizing that I was still better off staying with what I know best.

I must say that things really do happen for a reason and you never know what that is at the time. This realization became apparent when I went into teaching and needed thirty-nine college credits for my professional teaching license. I was able to use what I thought were wasted college courses to wipe out half the credits I needed. This was some thirty-plus years later.

In August 2012, I answered that ad and went for an interview with Porter and Chester Institute, who had just built a beautiful campus in Woburn, Massachusetts. After meeting with the Education Supervisor, Paul, and visiting the campus, I was excited. A day later, I received a call from Paul and was offered the position of automotive instructor. I was a bit taken aback by the pay offering, which was about twenty thousand dollars less than I was making. I really wanted to try it, and I knew that my employer would allow me to work weekends to supplement my income. I accepted the position and was eager to get started.

The building we moved into was empty, and I was in charge of ordering tools, equipment, lifts, benches, and all the necessary items to get an automotive shop up and running. I directed plumbers, electricians, painters, and equipment installers as to how I felt it should be laid out. After a few weeks of this, I had the task of making sense of their curriculum. I had never taught before so this was a bit challenging. What I was handed was all over the map in terms of organization and order of what should be taught when. I had all my years of experience behind me and was actually glad I had such input in this area. I was able to create a system that worked very well with some tweaking as I went along. I was now well on my way to becoming the teacher that I had signed up to become.

September 2012 was the next step in the process of becoming a teacher, students. It must be noted that I was a very shy kid. I grew up with eight siblings and had become a bit withdrawn as an adult, due to the turmoil that was my family. In school, I was always very nervous when called on to speak in class or partner up for projects. I had never considered a career with any type of speaking role, never mind public speaking. It should also be noted that

I have several teachers in my family who helped me mitigate my fears. The next step was presenting information to students.

My boss was aware that I came from industry and created an environment that allowed me to ease into my role as a teacher. We had week-long meetings and training with all the staff and had to present a short lesson at the end of the week that related to the subject matter in our field. I was nervous—very nervous. The woman presenting before me was hired as a dental instructor and went on for an hour and a half about brushing your teeth. I never thought that was possible! I must tell you that I was jumping out of my skin for two reasons, one being the anticipation of presenting and the other that I was so bored listening to her that I begged for the end.

I went next, standing in front of the 15 or so people. I presented information on automotive computers where I simplified their use making an analogy to how our brain interprets information from our vision, hearing, touch, taste, and smell to make decisions and act on that information. I asked the class how they would know if it was raining when they were driving their car down the street. They responded that they could see it, hear it, feel it, and sometimes smell it. *Exactly!* "Based on that information, what actions would you take if you were the driver and it started raining?" I asked. They said turn on the wipers and the lights. I went on to tell them that automotive computers work the same way. Based on inputs from sensors similar to your senses, the computer in the automobile will control fuel, turn on various devices, shift the transmission, and stop the car, just to name a few of its jobs. They loved that! I believe I hid my fear well, and everyone said I did a great job. The twenty minutes or so I presented marked a milestone in my new career. I have read that the more you teach, the more you start to believe that you are actually a teacher. It took me a while to overcome that fear of public speaking, but something did change in me that allowed me to concentrate on the material and not my own feelings. I became the lone night instructor and was responsible for all that went on at night, including classroom information and shop work. I designed tests, logically presented information, and graded all my students' assignments. I was responsible for their discipline and all matters that prevented them from graduating. We repaired customer vehicles, so I also booked work, wrote repair orders, directed student repairs, closed repair orders, and collected money for the repairs. If you are reading this and are on the fence about making a career change, push yourself outside your comfort zone. That's where growth takes place.

That job at Porter and Chester Institute went well for a couple of years until the money became an issue. The company had a wage freeze, so I had to move on. I now knew that I could teach and really enjoyed passing on my knowledge. The part-time job that I retained with the dealer turned back into a full-time job with much heartache.

Since that first teaching job, I had been actively seeking another full-time teaching position, which is hard to come by in the automotive sector. At the time, I also began pursuing my teaching credentials, so I would be ready to take on teaching in a public school system should the opportunity arise. Within two years, I had passed all the necessary tests for my preliminary license and was ready for the public schools. I transferred my previous college credits to a local college, so I could begin the process of obtaining a professional license. I had been on several job interviews with no luck. Most schools already had someone in mind and had to go through the process anyway. A job opened up at a local high school in the winter of 2015, and I went back after not being hired on the first two go rounds. I do believe everything happens for a reason. The first two teachers that were hired there had an awful time with the student population, and they were not backed up by the vocational director. The students walked all over them after seeing how things were run. They kicked in doors, broke windows, ruined some equipment, and stole tools. That is just what I heard about. Needless to say, their contracts were not renewed, and I was hired to start the following school year along with a new director. Turns out, the new director was all in and supported the program with great enthusiasm. The other instructor that had been there for thirty-three years had put in for retirement on my first year there. I was going to be in charge of running the program the following year. It took some doing, but I let the students know that they get what they give. They respect me, I respect them, simple as that. I let them know that I had a tremendous amount of experience and was looking to open career doors for them if they would just let me. The students began to respect the program and what we were doing for them. I had a great year that year and love teaching my craft.

After the turmoil the automotive program was in that first year I arrived, the kids appreciated what I brought to the school. One of the greatest highlights for me my first year was at the end of week one when I addressed the students about the upcoming year. I told them how excited I was to be working with them, helping pursue their career goals. When I was done, one of the students proclaimed that "the train is back on the tracks." That really touched me deep. Knowing the trouble they had the prior year made this statement all the more valuable to me. I feel like I am a great fit in this school and have a passion for what I do. I know that as long as I do my part to keep the tracks straight, this teaching train will remain on the tracks. I always let the kids know that if nothing else, you will know how your own vehicle operates, which will save you money, and who doesn't want to save money?

Every year brings new students with new personalities that create a classroom vibe or atmosphere. Some students make you laugh inside because what they say shouldn't be funny to the teacher. Other students make you laugh out loud because you just can't help it. They play off each other, and at times it is a challenge to get them back on track. One year, I had a class that was just fun and funny to be around. One student had a very infectious laugh—and he laughed a lot. Another student must have had a good-looking mother because the rest of the class always had something to say about her. Yet another student just had a way of saying the funniest things while keeping a straight face. Most days, I would have to teach from the back of the room or talk while walking away from them because I just could not get the smile off my face. I tried my best to keep my composure and get them to stop, especially when the jokes about the students' mothers were flying. But I'm sure I failed miserably. Some battles are not worth the fight. There are also some students who lose interest in the program halfway through the year, even though they must finish the year with you. Most of the time, that is not an easy situation. They can be disruptive and create a difficult learning environment. Many techniques are needed to teach kids these days due to all the stimulation they receive from the internet, video games, and cell phones, but I would not change my career choice.

I had one student, who was one of the most disruptive I have seen. He could not sit still or keep from talking. Prior teachers had thrown him out of class, and current teachers would tell me that he was intolerable. I was not going to treat him dismissively. Deep down, I knew he was a good kid who just had some things going on. One day, I gave a test, and one of the questions was: "What do I need to do to pass this course?" It was really a gimme question. This particular student made a point to show me as I was walking around that he had written "Don't be like me." I said that I didn't like that statement. I told him that if he had written, "Don't *act* like me," it would be more accurate. I told him he was a good kid, just a little fidgety. Well that went a long way, and he went on to be successful. He made it a point to come back to school to say hello from time to time. I believe that as long as students know you truly care about them and are competent, they will respect you and do the right thing. People say that teachers are lucky because we get summers off. Well, timewise that is true, but teachers need summers off for our mental recovery as well as professional organization to prepare to make the next year even better. We never really "punch out" of teaching.

There are also sad times. I recently had a student who was killed in a motorcycle accident at 17 years old. Very tragic! Other teachers I have talked to have had students killed by guns or in other accidents. It is not easy to deal with the death of a child who you are with every day for a year or more.

Since I started teaching in 2012, I picked up teaching nights for a couple of local tech schools. I began teaching a small engine course that I had designed. The program consists of three nights with two and a half hours per night for a total of seven and a half hours. The first five hours are in the classroom learning how the small engine works with presentations and show-and-tell pieces. I begin with basic engine construction and operation. I move on to carburation, then a basic electrical lesson, so I can explain ignition systems. I wrap up the classroom information talking about how to properly maintain a small engine. On our third night, we are in the shop repairing equipment that the students provide and disassembling equipment that I provide. The public eats this up. This class almost always fills up fast with a cap of fifteen students. I have a wide range of people sign up including women and children. Often no one wants to leave when the class is finished, and they just hang out talking and helping me clean up. These classes really make me love teaching. A local newspaper did an article on me and my teaching small engines. They came in, took pictures, and really talked me up. This made me feel like what I was doing was making a difference. I believe that if you are a homeowner, small engine skills should be a necessity.

Growing Up Automotive

From left: Mike Proulx, a software engineer from Pepperell, says he is not very handy but has lawn mowers and other tools that need fixing. Chance DiVianvittorio, a student at Littleton High School, wants to learn how to fix power tools to help around the house, and maybe help some neighbors. Jon Olden (Ayer), said, "I'm 52, and I want something to do on Monday nights."

Small engine repair a popular course
Teacher has been fixing small engines for over 35 years

By Jon Bishop
jbishop@nashobapub.com

WESTFORD — You want to learn how to repair a small engine? Then sign up for Nashoba Valley Technical High School's class on that very topic.

"We actually work in the automotive technology space," said Joann Sueltenfuss, community education director. "It's a hands-on class."

The class, which is seasonal, starts up again in April and will focus on spring equipment such as lawnmower engines. Interested people can sign up either online or by calling the school.

The last session had a group of about 8 to 12 people, which Sueltenfuss said is a good size.

The instructor, Charlie Rose, started at Nashoba Valley Technical High School last year. He's been fixing small engines, though, for over 35 years.

Teaching, he said, has "been working out well." He said he teaches for three weeks.

"I start out with a basic engine and how it operates," he said.

Then he moves into fuel delivery — from the gas tank to the carburetor.

"From the carburetion, I'll move onto electrical," he said. "We talk about a spark and how a spark is produced."

They'll talk magnetism, electricity, he said. And maintenance, too.

"The last evening, we are in the shop," he said. He'll have the students take apart a few lawnmowers.

He said he's had all sorts of people sign up for the class.

"I've had father and son pairs. I've had older gentlemen. I've had women," he said. "I have young kids, older kids. It runs the gamut, for sure."

His favorite part of teaching the class is when he knows the students understand it.

"You can see it on their faces," he said.

He also loves it when people thank him and tell him that they really enjoyed the class.

And what's great about his class is that it shows students how everything in the engine goes together. Some classes, he said, teach students how to take something apart and nothing more. But after taking his class, they'll learn how it all works.

"If you have a problem, you can fix it," he said. "You can understand what's going on with it."

That's why he encouraged people to sign up.

"If somebody's looking for a small engine class, I feel I'm doing it the best way," he said.

According to the Nashoba Valley Technical School District brochure, the small engine repair course begins April 6 and runs Mondays from 6 to 8:30 p.m. There are three sessions.

For more information on this and other courses, visit the Nashoba Tech community education website: http://www.nashobatech.net/community_ed.

Follow Jon on Twitter and Tout:
@JonBishopNP.

Local newspaper interview in 2015
Nashoba Valley Technical High School, Pepperell MA.

CHAPTER 11:

COWORKERS AND THE SHOPS

As I look back on my own long career, I believe automotive is the right choice for anyone interested in vehicles and who grew up like me, tinkering with anything they could get their hands on. If working with your hands and your brain appeals to you, this should be a top career consideration. Today's automobiles require much more brain power than they used to. Problem solving has become a large part of everyday repairs, and understanding how all the systems interact is a must. Picture yourself in a repair shop with all the necessary tools and equipment at your disposal. A vehicle has been towed to you from another shop with a unique problem that seems to stump others. You embrace the challenge, because you believe in yourself and your ability to figure it out with all the resources at your fingertips. You spend a couple hours going over all the basics to eliminate certain items and to move toward discovering possible causes. One by one, you test components and trace wiring. Then, bingo! You find a broken wire or a loose connector. You feel really good about yourself and smile as you complete the repair and return the vehicle to the happy customer. This is the satisfaction that comes with repairing automobiles and the motivation to move forward on the not-so-great days. Technology continues to march forward, so staying up-to-date is imperative in order to stay valuable to employers and make life in a shop just a little easier.

I have purchased houses, raised a family, and have always been free from worrying about money. This career, like most careers, has emotional peaks and valleys. When you land that first job, nerves may kick in. This is completely natural. Growth doesn't happen within your comfort zone. Anyone who is any good at anything started out knowing nothing. Remember that! I have always kept my eye on the big picture of why I took a certain job in the first place. In the early days, I saw a paycheck and never gave it a second thought. At the height of my career, probably about ten or so years in, I felt that I could fix anything and that I'd never have enough work to keep me busy.

My ability to move quickly from one automobile to the next and having a good idea of what was wrong gave me an edge. Jealousy was an obstacle at some of the places I've worked; I believe sometimes others have seen me as some kind of a threat to their career. I just wanted to be left alone to do my job and not get caught up in the drama of some shops. Negative talk and gossip can be poison to a shop. It's like a cancer that can spread out of control if it is allowed to grow. I feel that one of the best ways to control this cancer is to hold regular meetings. Often this will clear the air and allow people to vent about frustrations that build up within the festering mind of a technician. Another benefit of regular meetings is to share information on trends that some techs may be seeing and others may not. Most shops where I have worked do not hold regular meetings, but those that do benefit greatly. If you find this negative energy seeping into your workplace, steer clear of the talkers and do not get caught up in the atmosphere they are trying to create. The best thing to do is to let them know that if they have an issue with something in particular, they can either address it with the parties involved or walk out the door and find a better fit for themselves.

I remember one dealer where I worked with a shop foreman that was full of himself. He wasn't well-liked by his peers and felt that he knew pretty much everything there was to know about automobiles. This individual may have been in his late twenties or early thirties, and I came into that dealership with about 20 or so years in the field. I remained quiet and went about my business. I sensed the jealousy early on and never let it get to me. Although I was friendly with a few random people in the shop, as happens with most places, there were others that just seem to have a chip on their shoulder. One day I was fixing a vehicle that had no reverse lights. I knew the electrical circuit well and understood that there was a wire that went to the transmission to control the operation of the lights when placed in reverse. Well, when tracing out the reverse light circuit, I looked up and there was the shop foreman who brought the service manager over with him to try to make me look bad. The foreman asked me why I wasn't using the "transmission simulator." He turned toward the service manager to make sure he heard him. This simulator had many wires and switches that needed to be attached to the transmission. I knew that only one wire was involved, but apparently he didn't. The wire I was working with wrapped completely around the front of the car, and he saw me working up in the right-front area and shook his head as he said something to the service manager. It was difficult, but I swallowed my pride and stayed quiet. I found the broken wire right where I had been looking, and I repaired it. I cut that section of wire out and left it on the foreman's toolbox to send a clear message that I knew what I was doing. He never bothered me again.

Like I said, there are emotional highs and lows in this trade. When a particular job does not go as smoothly as it should and work starts backing up, you start to get really frustrated. Or if you have to fight a bolt or bracket to line up with the mounting hole, the blood pressure begins to rise. I've always tried to keep my composure but have witnessed firsthand the aggravation that boils over in some technicians. Wrenches thrown, drop lights smashed, and hammers used on innocent inanimate objects are just some of the outlets I have seen people use to vent the excess pressure. If you are in the vicinity of those outbursts, you also learn a whole new vocabulary of foul language. I have learned that if you just step away for a few minutes and return, the situation goes much smoother. The bolt usually just falls into the hole or the parts line up automatically. You cannot fight the negative energy; only change it to a positive force that is on your side.

Each shop has its own unique "feel." Some are quiet and everyone is busy working away, and some are noisy with radios going and guys laughing and joking. The loud shops are fun, but I have found that some guys take too much time out of their day horsing around which reduces their productivity. A lot of the shops I have encountered will have at least one complainer looking for anyone to take their side. These complainers will also reduce productivity, but they don't seem to care. They just want to be heard. I prefer to be friendly with everyone but only to a point, because working within a flat-rate pay system, time is money. Flat-rate, in case you don't know, is getting paid per job. Let's say a brake job calls for one and a half hours at the shop's labor rate. If the labor rate for the shop is one hundred dollars, then the customer will pay one hundred and fifty dollars in labor, and the technician will receive one and a half times their hourly rate for that brake job. There are set times for most repairs, but I have found a lot of shops abusing that system. The technician will tell the service writer what he wants to get paid for the repair, and many times, the writer will just charge the customer that amount without looking up what the actual time should be for that repair.

Within most shops there will be a senior tech or two who are the backbone of the establishment. A lot of shops will appoint one of these senior techs as shop foreman. They are seasoned professionals who understand how things work and get the majority of what we call the "problem vehicles." These are the vehicles that often require extra diagnostic time. If the shop uses a flat rate system, the A tech may get frustrated when they do not see hours adding up because of time lost. Any good service manager will see this dilemma and pay them accordingly. Technicians are hard to find, and shops know that. If you are a good employee and keep up on training, job security is one of the benefits. There will also be mid-level techs, or B techs, who produce a large portion of the income for the shops. They will take care of a lot of the fast-moving work and maintenance items. The rest of the shop employees will be made up of introductory-level technicians who take care of oil changes, lights, wipers, and other fast-moving items. They will park customers' vehicles, bring vehicles in and out for technicians, give customers rides, sweep floors, and empty barrels. Whatever they are asked to do, they will jump in and help. A good C-level tech is valuable to a shop as long as they see the value in what they do and don't complain about it. Shops are actively seeking out these C-level techs to train them to be their future A techs. They must learn to be patient and that's a difficult trait to find, especially in this age of kids who want instant gratification.

Behind the service desk is the service writer. They are responsible for making appointments, writing repair orders, and listening to customers' vehicle concerns. Often, they will dispatch work to technicians, write up estimates for technicians and sell work to customers. Service writers will usually work long hours and stay to collect money long after the technicians have left for the day. They could be assigned other tasks as well, such as distributing rental or loaner cars. A good service writer knows how to treat customers properly. They must sympathize with the vehicle and customer problems and also properly sell the needed work. I have worked with many service writers of varying talents and personalities and have worked with some at several different locations because either they changed jobs or I did. When you remain friends like that, you have another set of eyes and ears to give you insight into what it is like at their location. If a job opens up where they are, you now have an in if you were thinking of making a move. The same goes for them looking to make a move to your location.

Using a dispatcher to distribute work to technicians can be very valuable to a dealership. Most service departments do not use one, and they rely on the service writers to dispatch work directly to the technicians. When I have worked with good dispatchers, they have taken some of the load off both technicians and service writers. They write up estimates, check parts availability, check for any recalls or technical service bulletins, and perform their primary function of dispatching work to the proper technicians in a fair manner. In one dealership, I worked with the dispatcher to put together an "in-house" technical service binder. This binder contained unusual problems that we found within our own dealership. This gave us another resource when facing a challenge.

If a shop has a parts department, they are a valuable piece of the puzzle when it comes to repairing a vehicle properly. Having a part in stock, as opposed to having a technician stand around and wait while the part is ordered, makes life in a shop much easier. Also a smart counterperson knows not only to make sure they deliver the correct part to save time, but they know to hand out all the associated parts needed for a job, such as sealers and gaskets. Without this knowledge, the technician may have to make several trips back before completing the job.

Service managers are just that—managers of the service department. Sometimes, they have added responsibility of being a fixed operations manager, which means that they oversee all operations of what we call the back of the house which includes service, parts, clean-up, and body shop if there is one. I have had the good fortune to work with some of the best service managers in the industry, from whom I have learned a lot, and the misfortune of working with some of the worst. Some of the managers and writers have remained friends. And others I stay away from. Like most of the people you meet and work with, you know by just talking to them whether you will get along. You form an opinion from the start. Of course, there's always that middle ground where you just keep everything on a professional level. It will usually take some time to find out if they truly care about the customers they deal with or if they see them as a dollar sign, if they care about their shop employees or see them as a number that only produces income. The service manager really needs to be a motivator who keeps the balance between the general manager (who is all about shop production) and the happiness of the shop employees they manage. They need to be approachable and interact with everyone they encounter. Employees want to put in their best effort for the good managers, and really are unmotivated when working with the ones who seem to suck the motivation right out of the room. One possible career advancement for an automotive technician is moving up to the manager or assistant manager position. I have had this opportunity on more than one occasion but have always declined even though the pay would improve quite a bit. The stress accompanied with this position was not worth it to me.

One of my first encounters with a dealership-level service manager was John along with his foreman/service writer/brother Murray. They worked well together and kept the shop running smoothly. Murray was very knowledgeable about automobiles. When I had just started out in auto repair, I valued his opinion and took his advice on most things. John, on the other hand, really had no idea how automobiles operated but he was good at paperwork. As long as he stayed away from the shop operations, things ran well. John was one third owner of the dealership and lived close enough that he would go home for lunch every day. One sad day he did not return and we all wondered what had happened to him. It turns out that when he had gotten home that day, he found his wife had committed suicide. Those were sad days after that incident, and John was never the same.

At another dealership, the service manager was found dead from a heart attack with his key still in the dealership's door at 5 a.m. He was a very nice man who cared quite a bit about the shop and its employees. From what I have seen, running a service department can be very stressful. There is pressure to produce profit numbers for the service department, hiring and firing employees in order to have the best team in place, dealing with customers concerns which can get heated at times and making sure they are satisfied, working with the factory representatives in order to make sure paperwork is sent in correctly and coordinating with sales and parts for a smooth running operation.

In the best-run shops, the employees all work well together and interactions are effortless. Everyone knows their job, which makes for a pleasant day that seems to fly by. Conversely, in the worst-run shops, not many are happy, and the days drag on while watching the clock. When searching for a shop to settle in, there can be some telltale signs that will let you know how well the shop is run. For example, when driving up to the shop, is the parking organized and the signage clearly visible? Take a peek inside the shop. are the floors clean and free of clutter, and does everyone appear to be busy in their respective roles? At some point during the interview, did the service manager show you around to meet several members of their team?

Remember that you will be spending forty-plus hours a week in this environment, so choose wisely.

CONCLUSION:

So what will those of you thinking of a career in the automotive field be working on in the future? Fully electric and self-driving vehicles are what most experts see in the future. Some people do not understand that burning fossil fuels is still necessary to create the needed electricity to charge the batteries. Expanding clean energy such as solar and wind must be significantly increased to make automobile recharging stations feasible. Another problem is the disposal of all the battery packs after they have reached their ten-year lifespan. They can weigh upwards of five hundred pounds, and most are made of lithium ion. Currently only five percent of lithium ion batteries are recycled, which needs to be expanded if we are to continue the use of this technology. A lot of considerations need to be taken into account before electric automobiles become widespread—such as emergency measures for low battery power in an evacuation and providing power for those in remote areas with limited resources. In 2012, a major snowstorm in the U.K. brought many cities to a standstill due to electric vehicles running out of power and being abandoned. Should someone be injured in an accident, who would be at fault and how would alcohol play a role if the vehicle is fully autonomous? These are just some of the issues and questions that are being raised before this technology can become widely adopted.

The hydrogen fuel cell I spoke about earlier is thought to be one of the best chances for future fuel independence. The big problem at this stage of the technology is the amount of energy it takes to separate the hydrogen from the oxygen in water, or setting up the infrastructure needed to support these vehicles. Although there have been millions of miles on hydrogen test vehicles, we are just not there yet.

If you're just starting out as an employee in the automotive field, I don't see much change from internal combustion engines for some time. Technological changes are inevitable, and automotive technicians will adapt to technology as they always have. Vehicles will be powered in one way or another, and they will still have mechanical, electrical, and customer convenience systems that require regular maintenance and repair. Technician tool boxes may have more electrical test equipment in the future, but as far as I can see, we will always be in high demand. We need to stay up-to-date with the changing information in order to stay relevant. I hope that this book has inspired you and given you insight into what goes on in the day-to-day operations of the automotive service department.

ABOUT THE AUTHOR

Charlie Rose attended Medford Vocational Technical High School in the 1970s and has been in the automotive industry ever since. He is master and L1 certified with ASE and has been Master certified with Chrysler for over 25 years. He currently teaches automotive students at a local high school.

www.ingramcontent.com/pod-product-compliance
Lightning Source LLC
Chambersburg PA
CBHW070652220526
45466CB00001B/412